CHEMISTRY RESEARCH AND APPLICATIONS

STEARIC ACID

SYNTHESIS, PROPERTIES AND APPLICATIONS

CHEMISTRY RESEARCH AND APPLICATIONS

Additional books in this series can be found on Nova's website under the Series tab.

Additional e-books in this series can be found on Nova's website under the e-book tab.

CHEMISTRY RESEARCH AND APPLICATIONS

STEARIC ACID

SYNTHESIS, PROPERTIES AND APPLICATIONS

YUNFENG LIN
AND
QIANG PENG
EDITORS

New York

Library of Congress Cataloging-in-Publication Data

Stearic acid : synthesis, properties, and applications / editors, Yunfeng Lin and Qiang Peng (Sichuan University, P.R. China).
pages cm. -- (Chemistry research and applications)
Includes index.
ISBN 978-1-63463-172-3 (hardcover)
1. Stearic acid. 2. Esters. 3. Fatty acids in human nutrition. 4. Rubber chemicals. I. Lin, Yunfeng, editor. II. Peng, Qiang (Chemist) editor.
QD305.A2S814 2014
661'.8--dc23

2014037733

Published by Nova Science Publishers, Inc. † New York

CONTENTS

PREFACE

Stearic acid is an 18-carbon long-chain fatty acid, which is widely used in various fields, including rubber industry, food industry, cosmetic industry and pharmaceutical industry.

This book contains a collection of current knowledge about stearic acid in different fields with emphasis on its synthesis, properties and applications.

The book includes five chapters that mention the following topics: Introduction of chemosensory effects and the propertics and applications of stearic acid in food industry; Functions of stearic acid in rubber industry; Physicochemical properties and applications of stearic acid in biomedical sciences; Practical use as a lubricant in tablets direct compression; and the last topic is an overview on introduction and potential applications of deuterated hydroxystearic acids.

Yunfeng Lin and Qiang Peng
State Key Laboratory of Oral Disease,
West China School of Stomatology,
Sichuan University, China

In: Stearic Acid ISBN: 978-1-63463-172-3
Editors: Yunfeng Lin and Qiang Peng

Chapter 1

CHEMOSENSORY PROPERTIES OF STEARIC ACID

Gregory Smutzer[1*], and Judith C. Stull[2]

[1]Department of Biology, Temple University, Philadelphia, PA, US

[2]Department of Sociology, La Salle University, Philadelphia, PA, US

ABSTRACT

Dietary fats are composed of complex lipids that include long-chain fatty acids such as stearic acid. Stearic acid is a waxy solid at room temperature, and this eighteen-carbon saturated fat activates somatosensory pathways in the mammalian oral cavity. Studies on the chemosensory properties of stearic acid have been hampered by its low solubility in aqueous solutions, and its low volatility in the oral cavity. Nonetheless, recent studies do suggest that stearic acid activates chemosensory pathways in the oral cavity. Along with behavioral studies in rodents, recent human studies have demonstrated that stearic acid stimulates gustatory pathways in the oral cavity when textural cues are minimized. However, perceived taste intensity responses in humans are generally less than those obtained with eighteen-carbon *cis*-unsaturated fatty acids, or with short-chain fatty acids. Since stearic acid evokes minimal gustatory responses in the human oral cavity, this long-chain fatty acid has value as a masking agent in the manufacture of unpalatable drugs. In addition to its gustatory properties, humans can discriminate the

* Correspondence to: smutzerg@temple.edu (GS)

odor of stearic acid both orthonasally and retronasally. These results indicate that stearic acid also functions as an olfactory stimulus in the human oral cavity. In summary, accumulating evidence indicates that stearic acid stimulates gustatory, olfactory, and somatosensory pathways in the mammalian oral cavity. These important characteristics make stearic acid a useful chemosensory stimulus for modulating the flavor of food, for modulating the post-ingestive responses of fat-containing foods, and for masking the unpleasant taste of drugs.

INTRODUCTION

Humans consume a wide variety of foods that are rich in fats. These foods include hydrogenated oils such as palm and coconut oil, butter, animal fats such as lard and shortening, dark chocolate, fish oil, cheese, nuts, processed meats, and whipped cream. Many of these fat-containing foods are highly palatable to humans, and this appeal is caused by the various sensory systems that detect dietary lipids in the oral cavity [1]. This chemosensory response arises from the hydrolytic activity of triglycerides in the diet by lingual lipases [2], and by unesterified fatty acids that are normal constituents of many fat-containing foods [3]. The detection of fats in the oral cavity is thought to be caused by the integration of somatosensory (textural), olfactory, trigeminal, and gustatory cues [2, 4]. In addition, post-ingestive cues may further affect the perception of fats [2, 5-8].

Fatty acids are hydrolysis products of triacylglycerols, and these fats play an important role in human nutrition. Fats increase the flavor of food, which in turn increases food acceptance [9]. Hedonic ratings of fats often show a positive correlation with increasing body mass [10]. Due to their lack of hydration by water as well as their high number of covalent bonds, fats possess a high caloric content (~9 calories per gram). This high caloric content allows fats to be excellent sources of energy as they undergo oxidation in mitochondria. Thus, an understanding of fatty acid perception in the oral cavity is important for clarifying how individuals make choices regarding the type and amounts of fat-containing food they ingest [11].

Studies on fat chemoreception can determine why some individuals are better at regulating fat intake than others. These studies are important because the overconsumption of fats is linked to obesity [12], and perceived variations in fat perception may in turn increase the risk for cardiovascular disease and diabetes. Recent studies indicate that obese subjects exhibit a stronger preference for high fat and sweet foods than do lean subjects [5]. This preference may occur because obese subjects have a hyposensitivity to fatty acid perception in the oral cavity [2, 13]. This diminished fat taste sensitivity in obese individuals is thought to underlie a compensatory effect that causes an increased preference for, and intake of high fat foods. If so, then obese individuals are predicted to have higher lipid intake and greater body mass index than lean individuals [2, 8]. Thus, a better understanding of fat perception in the oral cavity has important implications to overall health, nutrition, and the control of obesity.

At present, the majority of evidence suggests that fats are primarily perceived in the oral cavity by their textural properties [10, 14]. Fatty acids whose hydrocarbon chains are longer than seven carbons are essentially insoluble in aqueous solutions and saliva. Therefore, undissolved or emulsified fats may stimulate a tactile response when these hydrocarbons come in contact with fat-sensing cells on the tongue and oral cavity. These oral somatosensory responses result in a perceived texture response [15, 16]. Secondly, an olfactory component is involved in oral fat perception [17, 18]. This olfactory component is caused by the volatile nature of fats in the oral cavity [18]. Thirdly, some (but not all) long-chain fatty acids activate the trigeminal nervous system in the human oral cavity. For example, linoleic acid depolarizes oral trigeminal neurons by releasing calcium from endoplasmic reticulum (ER) stores into the cytosol of activated sensory cells [19].

A fourth mechanism for oral fat perception is a gustatory component [20, 21]. A gustatory response to fatty acids is observed when textural cues are blocked by emulsifiers and thickening agents, and when nasal airflow is fully obstructed [20].

The contribution of a gustatory component by fatty acid stimuli in the oral cavity may be affected by their chain length [21], and by their degree of *cis*-unsaturation [2]. Subsequently, long-chain fatty acids such as stearic acid or linoleic acid likely exhibit a taste quality that is independent of textural and odorant properties [2, 4, 11, 20, 22]. A summary of fatty acids that are used in chemosensory studies is shown in Table 1.

The Oral Cavity and Lingual Lipases

Most dietary lipids are ingested as triglycerides, which are then degraded to free fatty acids for absorption across the intestinal epithelium. Lipases are soluble enzymes that hydrolyze triglycerides to glycerol and free fatty acids.

Table 1. Fatty Acids Used in Chemosensory Studies

Common Name	Number Carbons	Symbol [a]	Saturation	C-C Double Bonds	Melting Point	Molecular Weight	Solubility in water	Appearance
Caproic acid	6	C6:0	Saturated		-3.4	116.16	1.082 g/ 100 ml	Colorless oil
Lauric acid	12	C12:0	Saturated		44.2	200.32	0.006 g/ 100 ml	White powder
Stearic acid	18	C18:0	Saturated		69.6	284.48	0.0003g/ 100 ml	White powder
Elaidic acid	18	C18:1	Unsaturated	*trans*-9	45.0	282.46	Insoluble	White powder
Oleic acid	18	C18:1	Unsaturated	*cis*-9	13.4	282.46	Insoluble	Pale yellow liquid
Linoleic acid	18	C18:2	Unsaturated	*cis, cis* -9,12	-9.0	280.45	1.39×10^{-5}g/ 100 ml	Colorless oil
Linolenic acid	18	C18:3	Unsaturated	*cis,cis,cis* -9,12,15	-17.0	278.43	Insoluble	Light yellow oil

[a]Number of carbon atoms : Number of double bonds. Source: Voet, D. and Voet, J.G. Biochemistry, 2nd ed. John Wiley & Sons, New York. 1995.

Lipases are primarily produced in the pancreas, but these hydrolytic enzymes are also secreted into the mouth and stomach [23]. Pancreatic lipase hydrolyzes fatty acids from the C-1 and C-3 positions of triacylglycerols. In humans and rodents, Von Ebner's glands in the oral cavity secrete salivary lipase for fat digestion. This soluble enzyme catalyzes lipid breakdown in the absence of bile salts. Salivary lipase is also important for fat digestion in newborn humans since their pancreatic lipase output is minimal [24].

Rodents exhibit high levels of oral lipase activity, and lingual lipase is essential for the gustatory perception of fats in these animals [25]. For example, rodents prefer weak solutions of fatty acids over triglycerides or control solutions [26]. However, the amount of lipase activity in the human oral cavity is lower than in rodents. Several studies have identified pregastric lingual lipase activity in the human oral cavity. Katzenstein [27], Koebner [28], and Leone [29], all reported that lipase activity was present in human saliva. Up to 30% of fat can be hydrolyzed to partial glycerides and free fatty acids by lingual lipase within 20 minutes of ingestion [30]. More recently, Stewart and coworkers [31] reported that enzymatic activity of oral lipase could produce μmolar amounts of fatty acids, whose concentrations were sufficient to activate fatty acid receptors. However, human lingual lipase shows variable hydrolytic activity in various fat-containing foods. These recent results suggest that oral lipase play only a minor role in human oral fat detection [32].

Lingual lipase uses the catalytic triad of aspartatic acid, histidine, and serine to hydrolyze long-chain triglycerides into free fatty acids, and mono- and diacylglycerides [33]. However, lingual lipase has a pH optimum near 5, which is more acidic than the pH of saliva (saliva is unbuffered, and varies from pH 6.2 to 7.4). In addition, lipase activity continues in the stomach, where the secretion of alkaline pancreatic juice raises the pH of the digestion mixture so that hydrolysis of triacylglycerols can occur.

In humans, serum triglyceride levels are heritable and correlate with the risk of developing coronary heart disease [34]. In particular, the consumption of foods that are rich in esterified stearic acid is linked to coronary heart disease [35]. Foods such as meat, coconut oil and cocoa butter are rich in stearic acid content [32, 36]. The salivary content of several commonly consumed fat-rich foods such as shredded coconut, meat, walnut, almond, almond butter, and olive oil was recently examined by GC-mass spectrometry [3]. Stearic acid was one of four predominant fatty acids detected in saliva, with concentrations up to 60 μmolar. Interestingly, the hydrolysis of olive oil resulted in the greatest release of stearic acid into the oral cavity [3]. These results indicate that long-chain fatty acids such as stearic acid are of sufficient concentrations to initiate gustatory signaling pathways in the human oral cavity when high-fat foods are ingested.

Orlistat (tetrahydrolipstatin) is a reversible inhibitor of oral, gastric, and pancreatic lipases. This inhibitor forms an ester linkage with serine (serine-152) in the active site of the lipase so that a decrease in the rate of lipid hydrolysis will occur [37]. Orlistat decreases the oral taste sensitivity to both oleic acid and triolein in obese subjects [38]. These results further suggest a role for lingual lipase in oral fat hydrolysis and detection. However, a conclusive role for human lingual lipase in oral fatty acid hydrolysis and detection still remains unclear at the present time [39].

Fatty Acid Taste Perception in the Oral Cavity

Humans detect at least five primary taste stimuli, which include sweet, sour, salty, bitter, and *umami* taste. Representative taste stimuli for the five primary taste qualities consist of polar molecules that are generally presented as aqueous solutions to subjects for psychophysical studies. Medium- and long-chain fatty acids are non-polar molecules that do not readily dissolve in water. This low solubility in water increases the difficulty of human psychophysical studies because sufficient amounts of medium- and long-chain fatty acid stimuli are difficult to present to subjects at amounts that elicit a strong chemosensory response. A second obstacle in identifying fats as true gustatory stimuli is the difficulty in fully excluding the contributions of other sensory systems [2, 20], especially textural cues in the oral cavity.

Recent evidence from both animal and human studies suggests that oral detection of fatty acids occurs via a gustatory pathway [3]. As opposed to the five primary taste stimuli, fat taste appears to have no discrete taste quality [40]. However, humans do possess a dietary requirement for fatty acids. Essential fatty acids such as linoleic acid are important for plasma membrane formation, and for the absorption of fat-soluble vitamins [41]. Due to the importance of fatty acids in the diet, the generation of a specific taste quality by fatty acids is a reasonable hypothesis.

Long-chain fatty acids such as stearic acid or linoleic acid may activate signaling systems in the oral cavity, intestines, and CNS of rodents [42, 43]. Based on animal studies, several mechanisms have been proposed for chemosensory transduction of long-chain fatty acids in the oral cavity. One mechanism involves the inhibition of a delayed rectifying K^+ channel by long-chain fatty acids in isolated taste receptor cells [44, 45].

Alternatively, fat taste may occur by interactions with cluster of differentiation-36 (CD36) protein [46, 47]. CD36 is a scavenger receptor that mediates lipid trafficking in a variety of cell types [48]. This receptor-like glycoprotein binds to both saturated and unsaturated long-chain fatty acids with an affinity in the nanomolar range [49]. CD36 is a multifunctional protein that also binds modified phospholipids and thrombospondins in other cell types [50]. This cell surface glycoprotein may function as either a fatty acid transport molecule, or as an oral fat receptor. In addition, this receptor mediates fatty acid-induced release of gut peptides in rodents [51]. As shown in Figure 1, fatty acid binding may occur at or near a positively charged lysine residue that is located within a hydrophobic pocket of CD36. This pocket is located on the extracellular surface of this protein [52]. CD36 mediates cell signaling by increasing intracellular calcium by emptying ER calcium stores, which in turn open plasma membrane store operated channels [53, 54]. Notably, CD36 protein localizes to the apical region of cells in taste buds of foliate and circumvallate papillae of rats [55]. CD36 also localizes to the human GI tract, which further supports a physiological role for this receptor in the absorption of dietary lipids [56].

A preference for linoleic acid in both obesity-prone and obesity-resistant rats is attenuated by a reduction of CD36 on the tongues of these rodents [57]. Both the *CD36* gene and lingual lipase are thought to influence oral sensitivity to fat in obese subjects [38]. Finally, several single nucleotide polymorphisms in the human CD36 gene have been identified that successfully predict oral responses to fat [38, 58]. Taken together, these results suggest that CD36 protein functions as a fatty acid chemoreceptor in the human and rodent oral cavity.

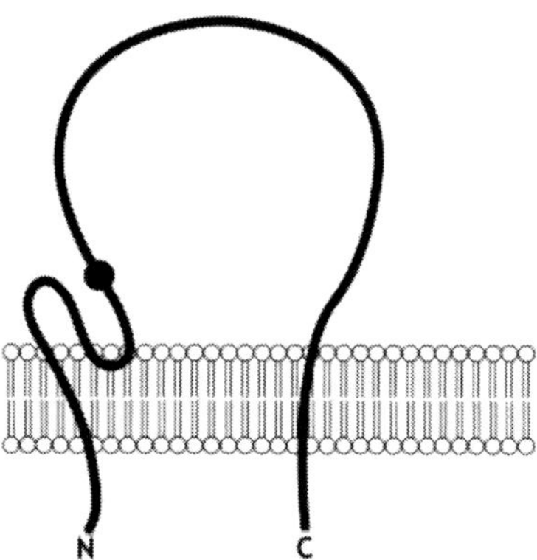

Figure 1. Schematic of CD36. This glycoprotein is predicted to contain two transmembrane domains, and a hydrophobic pocket for binding long-chain fatty acids. Lysine-164 (filled black circle) is predicted to exhibit intermediate solvent accessibility, and localizes to the hydrophobic pocket that contains the putative binding domain for long-chain fatty acids. N and C represent the amino and carboxyl termini of the protein. Adapted from Kuda et al. [52]. Image courtesy of Eric B. Tran.

Another view is that the G-protein coupled receptors GPR40 and GPR120 mediate taste preferences for medium and long-chain fats via the GTP binding proteins $G\alpha_q/G\alpha_{11}$ [59, 60]. As opposed to CD36, these receptors possess a low affinity for fatty acids [49]. Finally, transient receptor potential cation channel, subfamily M, member 5 [TRPM5] receptors may play an important role in fatty acid perception in rodents since inactivation of the *TRPM5* gene abolished preferences for fats in mice [61, 62].

In contrast, some evidence suggests that fatty acid perception in the oral cavity is not receptor mediated. Fatty acid chemoreception may occur by passive diffusion across taste cell membranes [63, 64]. Long-chain fatty acids could form micelles in salivary fluid at amounts above their critical micelle concentration. In addition, fatty acid monomers could be in thermodynamic equilibrium with fatty acid micelles in saliva. These micelles could enhance chemosensory responses in the oral cavity by endocytosis across plasma membranes of oral taste receptor cells. In addition, fatty acid monomers could directly diffuse across the plasma membrane of fat-sensing cells. Once inside the cytoplasm, these fats could activate cell signaling pathways and initiate a chemosensory response.

Stearic acid is a white, waxy solid at room temperature. Since stearic acid contains a polar carboxyl group at C-1 and a non-polar hydrocarbon tail, this fatty acid is an amphipathic molecule that may form micelles in aqueous solutions at temperatures above its melting point. Based on structural analogy with other carboxylic acids, the estimated pKa of stearic acid is ~5.0 [65]. Since saliva has a pH above the pKa of stearic acid, this fat is predicted to exist almost entirely in its anion form when solubilized. The anionic form of stearic acid is unlikely to cross the hydrophobic region of a plasma membrane by simple diffusion. In addition, the relatively long hydrocarbon chain of stearic acid might decrease its diffusion rate across the lipid bilayer of fat-sensing cells. However, hydrophobic compounds such as capsaicin (which does not contain a polar carboxyl group) do diffuse across plasma membranes before binding to TRPV1 receptors at ligand binding sites that face the cytosol. Capsaicin requires 20-30 seconds for a maximal chemosensory response [66], which appears to be longer than the time required for a fatty acid chemosensory response in the oral cavity. A primary role for diffusion of fatty acids across plasma membranes in order to activate fat-sensing cells in the oral cavity needs further examination.

The activation of taste receptor cells by fatty acids in turn stimulates the lingual nerve (branch of the mandibular nerve), which contains fibers of the chorda tympani (CT) branch of the facial nerve. The CT nerve detects taste stimuli in the anterior two-third portion of the tongue. In addition, the glossopharyngeal nerve detects taste stimuli in the posterior third of the tongue [67]. These two nerves carry sensory information to the nucleus of the solitary tract, a vertical column of grey matter that is embedded in the medulla oblongata [68]. For pleasurable taste stimuli, projections from the nucleus of the solitary tract terminate in the postrolandic sensory cortex (area 43 of the parietal lobe). These projections are thought to cause the release of dopamine in the ventral striatum, a region of the forebrain that is involved in processing the hedonic quality of food [42]. The release of dopamine could cause a pleasurable response [69], which in turn might stimulate further intake of fat by positive feedback.

The cephalic phase of gastric secretion occurs before food enters the stomach, and is thought to occur when food enters the oral cavity. This response prepares the gastrointestinal (GI) tract for the optimal processing and absorption of ingested foods [40]. Enzymes for the hydrolysis and absorption of fat in the GI tract may become activated during the ingestion of fatty foods. As opposed to *cis*-unsaturated fats such as oleic, linoleic, and linolenic acids, wild-type mice with esophageal ligation do not show increased pancreatic and bile duct secretion after exposure to stearic acid [70]. Thus, stearic acid appears to have less of an effect on the post-ingestive satiety of fats when compared to long-chain *cis*-unsaturated fatty acids [71].

Activation of the Human Olfactory System by Stearic Acid

The focus of this chapter is on the chemosensory properties of stearic acid in the human oral cavity. Stearic acid is an eighteen-carbon, saturated fatty acid. As with other long-chain fatty acids, stearic acid is thought to be detected by multiple sensory systems in the human oral cavity [4, 7, 8, 11, 20, 22].

Recent animal studies have demonstrated that fatty acids do possess an olfactory component [18, 72]. For oral chemosensory stimuli, olfactory cues can occur by either an orthonasal or a retronasal route. Stearic acid has a vapor pressure of ~0.8 mm of mercury at 180 °C. [73], and may become volatile in the oral cavity at body temperature. For example, olfactory nerve sectioning in mice eliminates preferences for high-fat foods, and recovery of neuronal connections results in the re-establishment of these preferences [18]. In humans, Chalé-Rush et al. [4] reported both orthonasal and retronasal olfactory responses to stearic acid, with retronasal olfaction showing the highest detection thresholds. At suprathreshold concentrations, these researchers reported that vapor-phase stearic acid could be discriminated from control mineral oil vapor by both orthonasal and retronasal routes. These results suggest that stearic acid elicits an olfactory response in the human oral cavity [4].

Activation of the Human Trigeminal Sensory System by Stearic Acid

Trigeminal somatosensory neurons detect thermal, mechanical, and chemical stimuli in the oral cavity [74]. By fully blocking nasal airflow, "oral cavity only" detection of vapor phase stearic acid was examined [75]. The purpose of this study was to establish whether this saturated fat could function as a trigeminal stimulus in the oral cavity. Vapor phase stearic acid could not be discriminated from controls, and suggested that stearic acid did not activate the oral cavity component of the human trigeminal sensory system [4, 8, 75].

Activation of the Human Gustatory System by Stearic Acid

Mounting evidence suggests that humans may detect the taste of free fatty acids [20]. This taste quality is primarily ascribed to the detection of long-chain fatty acids that vary in chain length and saturation [2, 4, 20, 22, 38, 76].

A major roadblock in identifying gustatory properties of stearic acid is the absence of a delivery system that can present high concentrations or amounts of hydrophobic stimuli in the absence of tactile (and olfactory) cues. For example, Chalé-Rush et al. [4, 22] and Mattes [2] found that the detection of smell and taste thresholds of fatty acid stimuli differed with presentation conditions. This absence of a suitable delivery system currently limits suprathreshold studies on fatty acid chemosensation in the oral cavity.

Since stearic acid is a solid that is virtually insoluble in saliva, this lipid is predicted to yield a minimal taste response when compared to long-chain fatty acids that are oils at physiological temperature. Saturated fats such as stearic acid produce insoluble mixtures in water that may cause both a tactile and a chemosensory response. However, hydrophobic chemosensory stimuli can be presented to subjects as complex emulsions of gum acacia, EDTA, water, and mineral oil that mask the viscosity of fatty acid stimuli [2, 4, 21, 22, 77]. These emulsions may trigger a tactile response on the tongue surface, and may show heterogeneity in the distribution and amounts of fatty acid that are suspended in the emulsion.

Stearic acid was successfully delivered to the human oral cavity by emulsions at 67-69 ^{o}C where this stimulus is in liquid form [4, 22]. Detection thresholds were identified by orthonasal olfaction, retronasal olfaction, gustation, and a multimodal presentation where the lipid emulsion was placed in the oral cavity in the absence of nose clips [4]. Although measured at different temperatures, intensity responses for stearic acid were similar to the 18-carbon *cis-* unsaturated fatty acids linoleic and oleic acid. Oral detection thresholds for stearic acid in the human oral cavity with emulsions, yielded thresholds near 0.032% (w/v) [4, 22]. In addition, most study participants were able to detect stearic acid in the oral cavity [4, 22].

Oral detection thresholds were also obtained by an ascending, three-alternative, forced-choice procedure for three saturated fatty acids that varied in chain length, and one long-chain *cis*-unsaturated fatty acid [2]. Although textural cues were not eliminated, stearic acid emulsions in the oral cavity yielded a mean detection threshold near 0.03% w/v [2]. Similar threshold results were obtained with and without nasal airflow. In addition, the mean detection threshold for stearic acid in this population was statistically similar to thresholds for caproic, lauric, and linoleic acids. These results suggest that humans can perceive short-, medium- and long-chain fatty acids in the oral cavity with detection thresholds that do not appear to correlate with the length of the fatty acid chain.

Median threshold concentrations for stearic acid at circumvallate, foliate, and fungiform papillae on the tongue surface of humans have also been identified. Detection thresholds for stearic acid in these three tongue regions in either the presence or absence of nasal airflow covered four orders of magnitude, with ranges that varied from 0.0003 to 2.8% w/w [78]. Detection thresholds showed nearly identical concentrations for vallate, foliate, and fungiform papillae, with means near 0.002% w/w. In addition, this study found that the ability of participants to monotonically rate the intensity of stearic acid at vallate, foliate, and fungiform papillae as the amount of stimulus increased, was lower for stearic acid than for shorter chain fatty acids such as caproic acid or lauric acid [78]. These results suggested that humans are capable of scaling the intensity of graded fatty acid concentrations [78].

More recently, stearic acid has been incorporated into edible strips that contain the polymers pullulan and hydroxypropyl methylcellulose [11]. With this delivery method, control strips and strips with taste stimuli yield similar tactile responses. In the presence of nasal airflow, Ebba et al. [11] reported that

All taste quality responses were normalized to 100%. "Other" taste represents a response that is not perceived as sweet, sour, salty, or bitter. Subjects who responded "other" were further asked to use a word description to identify the taste quality of fatty acid taste strips. Data in parenthesis in column seven represent the percentage of "other" tasters who gave a taste quality response of fatty/oily/waxy/sunflower seed taste. Data for stearic acid is from 30 subjects, and includes one light smoker (< 10 cigarettes per day). Modified from Ebba et al. [11].

stearic acid yielded a chemosensory response that was in the barely detectable range when measured with the general Labeled Magnitude Scale. Intensity values for stearic acid were lower than those obtained with the *cis*-unsaturated fat linoleic acid [11]. In addition, one-sixth of test subjects reported a response of no taste at amounts up to 1.7 μmoles of stearic acid. For individuals who reported an intensity response for stearic acid, approximately one-fifth of respondents described the flavor as a fatty/oily/waxy/sunflower seed taste (See Table 2). As the amount of stearic acid increased in the strips, a greater proportion of respondents described the taste quality of stearic acid as bitter. Intensity values for stearic acid inversely correlated with hedonic responses. Finally, the ability to perceive the bitter taste of *n*-propylthiouracil did not affect the ability to detect stearic acid perception in the oral cavity [11].

Table 2. Taste Quality Responses to Various Amounts of Stearic Acid

µmol of Stearic acid	Sweet	Sour	Salty	Bitter	No discernible taste	Other
0.0	0	0	1	7	80	11 [46]
1.1	0	2	0	7	31	60 [48]
1.3	0	5	0	5	21	69 [79]
1.5	0	5	0	12	14	69 [48]
1.7	0	0	0	31	41	29 [67]

Drug Delivery and Stearic Acid Microspheres

The oral administration of therapeutic agents is the most common method of drug delivery to the human body, and is the preferred route for non-invasive drugs [79]. Over the last thirty-five years, polymer-based microspheres have been developed as a vehicle for microencapsulating pharmaceuticals for use in humans [80]. These biocompatible microspheres form small rounded particles in aqueous solutions that vary from 1 to 1000 µmeters in diameter, and show promise as potential drug delivery systems [81, 82]. Microspheres also permit the sustained release of drugs so that the therapeutic action of the encapsulated medication occurs over a longer period of time [83, 84]. In addition, microspheres can be coated with molecules that mask the bitter taste of encapsulated pharmaceuticals [85-87].

Stearic acid is a straight chain hydrocarbon that contains no *cis* double bonds, and this fatty acid can pack tightly onto the surface of microspheres. Stearic acid is also a hydrophobic molecule with a melting point that is well above body temperature. In addition, stearic acid exhibits a minimal taste response in the oral cavity [11]. Taken together, these characteristics favor the use of stearic acid as a taste masking agent for bitter-tasting drugs [83], and as a possible agent for regulating drug release from microspheres [84].

Drug release from stearic acid-coated microspheres may occur by diffusion when microspheres come in contact with aqueous fluids in the GI tract. This release may occur when water diffuses into the interior of the microsphere, and allows the drug to diffuse across the release coat of the microsphere to the exterior [82]. In addition, stearic acid may act as an emulgent that regulates the rate of drug release from microspheres [83]. After ingestion, the drug will disperse and dissolve upon contact with GI fluid, possibly by interactions between stearic acid and the suspension media [85].

Recently, stearic acid has been used to coat the surface of ibuprofen-containing microspheres [84]. In these preparations, stearic acid only partially delayed the release of ibuprofen when microspheres were suspended in a variety of buffer conditions. A scanning electron micrograph image of representative stearic acid microspheres that contain ibuprofen is shown in Figure 2.

In addition, stearic acid shows considerable promise in masking the bitter taste of drugs [85-87]. This approach has important applications for administering bitter-tasting drugs for human consumption, and for pets and agricultural animals. In particular, lipid coated microsphere suspensions should increase the palatability of drugs for children [88-90]. Stearic acid coated microspheres can mask the unpalatable taste of antibiotics [85-87]. For example, stearic acid has been formulated with the cephalosporin antibiotic cefuroxime axetil (CEFTIN) where drug particles were coated by a spray chilling process that produced stearic acid-coated axetil. These preparations were then prepared as suspensions for drug delivery. Stearic acid successfully masked the bitter taste of cefuroxime axetil in these microspheres [85-87].

Fatty acid microspheres also show promise for the oral administration of a variety of particulate drugs. Carnauba wax (palm wax) and Compritol® 888 ATO (glyceryl behenate) are useful lipid matrices that can be coupled to varying concentrations of stearic acid [91]. This lipid assembly stimulates the release of drug when stearic acid becomes ionized at basic pH. In addition, stearic acid-chitosan-gelatin blended microspheres [91] have been prepared by ultrasound-assisted atomization of molten lipid dispersions at weight ratios of 10% or 30% (w/w) stearic acid. Stearic acid remained on the exterior of the microsphere, and coated its surface. At basic pH, the dissolution of stearic acid decreased the packing ratio of lipids (width of the carboxyl headgroup compared to width of fatty acid tail) in microspheres. This decreased packing ratio may have been caused by the charged carboxyl group of the fatty acid which enhanced the dissolution of the microsphere, and stimulated release of the encapsulated drug.

Figure 2. Scanning electron micrograph image of stearic acid microspheres prepared by immersion in pH 8.0 sodium phosphate buffer at 37°C according to Waters & Pavlakis [84].

Future Directions

One major limitation of taste studies in the human oral cavity is the lack of a standardized test that can be used for both hydrophilic and hydrophobic taste stimuli. In addition, long-chain saturated fats (such as stearic acid) often yield minimal gustatory responses in humans. This minimal response complicates the measurement of gustatory responses of long-chain saturated fats in the oral cavity. In order to advance the growing field of fatty acid chemoreception, improved delivery methods must be developed that allow sufficient amounts of fatty acid to be presented to subjects for suprathreshold measurements while minimizing the tactile response of these stimuli. Under these conditions, a true gustatory response for fatty acids can be systematically examined in the human oral cavity. These studies could further identify whether some individuals exhibit taste blindness to one or more fatty acid stimuli [92], and if a unique taste quality can be attributed to fatty acids.

Secondly, a plasma membrane receptor or carrier for activation of taste receptor cells by fatty acid stimuli must be conclusively identified and characterized. The identification of a fat taste receptor could then determine receptor localization and receptor density in the oral cavity, which in turn could clarify its role in regulating fat consumption in humans. These studies could further identify genetic components that may regulate fat consumption in humans [93].

Thirdly, animal studies suggest that fatty acid taste in the oral cavity enhances other primary taste stimuli [44]. In humans, hedonic preferences for sweet taste stimuli are often enhanced by raising the fat content of food [15, 94, 95]. Foods with a high fat and sugar content generally promote overeating in humans [9]. Other studies suggest that the component taste qualities of mixtures generally remain separate in both humans and rodents [96]. Nonetheless, a systematic examination of fatty acid-sweet taste mixtures could explain why these mixtures enhance food consumption in humans.

Finally, the physical properties of each fatty acid are critical for their detection by fat-sensitive cells in the oral cavity. The effect of carbon chain length, degree of unsaturation, orientation of carbon-carbon double bonds, and physical state of fatty acids on chemosensation and somatosensation in the oral cavity must be fully clarified. Most importantly, the physical properties of fatty acids affect their solubility in saliva, which in turn affect chemosensory responses for individual fatty acids in the oral cavity. Recent studies do suggest that carbon chain length affects oral sensitivities to fatty acids [21].

CONCLUSION

Recent evidence suggests that the oral detection of fatty acids such as stearic acid do occur in the human and rodent oral cavity, and stimulate a fatty acid taste response. Both animal and human studies indicate that stearic acid elicits a small to moderate taste response in the oral cavity. The hedonic appeal of fats may be enhanced especially when consumed with carbohydrates. Future studies will determine whether long-chain fatty acids such as stearic acid or linoleic acid represent a primary taste stimulus in the human oral cavity. Finally, improved delivery methods for saturated and unsaturated fatty acids for psychophysical studies will stimulate the advancement of this important field of study.

Acknowledgments

This work was supported by a Temple University Faculty Senate Seed Money Grant. The authors thank Susan E. Coldwell, Leonard X. Finegold, and Joseph T. Tran for their valuable assistance.

References

[1] DiPatrizio, N.V., Joslin, A., Jung, K.M. & Piomelli, D. Endocannabinoid signaling in the gut mediates preference for dietary unsaturated fats. *FASEB J.* 2013, 27: 2513-2520.

[2] Mattes, R.D. Oral detection of short-, medium-, and long-chain free fatty acids in humans. *Chem. Senses* 2009, 34: 145-150.

[3] Kulkarni, B.V. & Mattes, R.D. Evidence for presence of nonesterified fatty acids as potential gustatory signaling molecules in humans. *Chem. Senses* 2013, 38: 119-127.

[4] Chalé-Rush, A., Burgess, J.R. & Mattes, R.D. Multiple routes of chemosensitivity to free fatty acids in humans. *Am. J. Physiol. Gastrointest. Liver Physiol.* 2007, 292: G1206-G1212.

[5] Rolls, E.T., Critchley, H.D., Browning, A.S., Hernadi, I. & Lenard, L. Responses to the sensory properties of fat of neurons in the primate orbitofrontal cortex. *J. Neurosci.* 1999, 19: 1532–1540.

[6] Kirkham, T.C. & Tucci, S.A. Endocannabinoids in appetite control and the treatment of obesity. *CNS Neurol. Disord. Drug Targets* 2006, 5: 272-292.

[7] Rolls, E.T. Neural representation of fat texture in the mouth. In: Montmayeur, J.-P. & le Coutre, J., Editors. Fat detection: Taste, texture, and post ingestive effects. Boca Raton, FL. CRC Press. 2010, pp. 197-224.

[8] Bolton, B. & Halpern, B.P. Orthonasal and retronasal but not oral-cavity-only discrimination of vapor-phase fatty acids. *Chem. Senses* 2010, 35: 229-238.

[9] Drewnowski, A. & Almiron-Roig, E. Human perceptions and preferences for fat-rich Foods. In: Montmayeur, J.-P. and le Coutre J., Editors. Fat detection: Taste, texture, and post ingestive effects. Boca Raton, FL: CRC Press. 2010, pp. 265-289.

[10] Mattes, R.D. Fat taste and lipid metabolism in humans. *Physiol. Behav.* 2005, 86: 691-697.

[11] Ebba, S., Abarintos, R.A., Kim, D.G., Tiyouh, M., Stull, J.C., Movalia, A. & Smutzer, G. The examination of fatty acid taste with edible strips. *Physiol. Behav.* 2012, 106: 579-586.

[12] Drewnowski, A., Kurth, C., Holden-Wiltse, J. & Saari, J. Food preferences in human obesity: carbohydrates versus fats. *Appetite* 1992, 18: 207–221.

[13] Stewart, J.E., Seimon, R.V., Otto, B., Keast, R.S., Clifton, P.M. & Feinle-Bisset, C. Marked differences in gustatory and gastrointestinal sensitivity to oleic acid between lean and obese men. *Am. J. Clin. Nutr.* 2011, 93*:* 703–711.

[14] Rolls, E.T., Verhagen, J.V. & Kadohisa, M. Representations of the texture of food in the primate orbitofrontal cortex: neurons responding to viscosity, grittiness, and capsaicin. *J. Neurophysiol.* 2003, 90: 3711–3724.

[15] Drewnowski, A., Shrager, E.E., Lipsky, C., Stellar, E. & Greenwood, M.R. Sugar and fat: sensory and hedonic evaluation of liquid and solid foods. *Physiol. Behav.* 1989, 45: 177-183.

[16] Drewnowski, A. Why do we like fat? *J. Am. Diet. Assoc.* 1997, 97 (7 Suppl.): S58-S62.

[17] Ramirez, I. Role of olfaction in starch and oil preference. *Am. J. Physiol.* 1993, 265 (6 Pt 2): R1404-R1409.

[18] Kinney, N.E. & Antill, R.W. Role of olfaction in the formation of preference for high-fat foods in mice. *Physiol. Behav.* 1996, 59: 475–478.

[19] Yu, T., Shah, B.P., Hansen, D.R., Park-York, M. & Gilbertson, T.A. Activation of oral trigeminal neurons by fatty acids is dependent upon intracellular calcium. *Pflugers Arch.* 2012, 464: 227-237.

[20] Mattes, R.D. Is there a fatty acid taste? *Ann. Rev. Nutr.* 2009, 29: 305-327.

[21] Running, C.A. & Mattes, R.D. Different oral sensitivities to and sensations of short-, medium-, and long-chain fatty acids in humans. *Am. J. Physiol. Gastrointest. Liver Physiol.* 2014, 307: G381-G389.

[22] Chalé-Rush, A., Burgess, J.R. & Mattes, R.D. Evidence for human orosensory (taste?) sensitivity to free fatty acids. *Chem. Senses* 2007, 32: 423-431.

[23] Nelson, J.H., Jensen, R.G. & Pitas, R.E. Pregastric esterase and other oral lipases – a review. *J. Dairy Sci.* 1977, 60: 327-362.

[24] Hamosh, M. & Hamosh, P. Development of digestive enzyme secretion. In: Sanderson, I.R. and Walker, W.A., Editors. Development of the gastrointestinal tract, Vol. 1. Hamilton, ON. B.C. Decker, Inc. 1999, pp. 261-278.

[25] Kawai, T. & Fushiki, T. Importance of lipolysis in oral cavity for orosensory detection of fat. *Am. J. Physiol. Regul. Integr. Comp. Physiol.* 2003, 285: R447–R454.

[26] Tsuruta, M., Kawada, T., Fukuwatari, T. & Fushiki, T. The orosensory recognition of long-chain fatty acids in rats. *Physiol. Behav.* 1999, 66: 285-288.

[27] Katzenstein, M. Untersuchungen über speichellipase. *Z. Ges. Exp. Med.* 1930, 69: 179-192.

[28] Koebner, H. Untersuchungen über speichellipase. *Z. Ges. Exp. Med.* 1931, 76: 792-803.

[29] Leone, V. Clinico-statistical study of salivary lipase. *Minerva Stomatol.* 1955, 4: 18-23.

[30] Shepherd, R. W. & Cleghorn, G. J. Disturbances of gastrointestinal disturbances in CF. In: Cystic fibrosis: Nutritional and intestinal disorders. Boca Raton, FL: CRC Press. 1989, p. 83.

[31] Stewart, J.E., Feinle-Bisset, C., Golding, M., Delahunty, C., Clifton, P.M. & Keast, R.S. Oral sensitivity to fatty acids, food consumption and BMI in human subjects. *Br. J. Nutr.* 2010, 104: 145-152.

[32] Kulkarni, B.V. & Mattes, R.D. Lingual lipase activity in the orosensory detection of fat by humans. *Am. J. Physiol. Regul. Integr. Comp. Physiol.* 2014, 306: R879-R885.

[33] Hermetter, A. Triglyceride lipase assays based on a novel fluorogenic alkyldiacyl glycerol substrate. In: Doolittle, M., and Reue, K., Editors. Lipase and phospholipase protocols. Totawa, NJ. Humana Press, Inc. 1999, pp. 19-30.

[34] The TG and HDL working group of the exome sequencing project, National Heart, Lung & Blood Institute. Loss-of-function mutations in APOC3, triglycerides, and coronary disease. *New Engl. J. Med.* 2014, 371: 22-31.

[35] Berry, S.E. Triacylglycerol structure and interesterification of palmitic and stearic acid-rich fats: an overview and implications for cardiovascular disease. *Nutr. Res. Rev.* 2009, 22: 3-17.

[36] Wood, J.D., Richardson, R.I., Nute, G.R., Fisher, A.V., Campo, M.M., Kasapidou, E., Sheard, P.R. & Enser, M. Effects of fatty acids on meat quality: a review. *Meat Sci.* 2003, 66: 21-32.

[37] Guerciolini, R. Mode of action of orlistat. *Int. J. Obes. Relat. Metab. Disord.* 1997, 21(Suppl. 3): S12-S23.
[38] Pepino, M.Y., Love-Gregory, L., Klein, S. & Abumrad, N.A. The fatty acid translocase gene CD36 and lingual lipase influence oral sensitivity to fat in obese subjects. *J. Lipid Res.* 2012, 53: 561–566.
[39] Schiffman, S.S., Graham, B.G., Sattely-Miller, E.A. & Warwick, Z.S. Orosensory perception of dietary fat. *Curr. Dir. Psychol. Sci.* 1998, 7: 137–143.
[40] Newman, L., Haryono, R. & Keast, R. Functionality of fatty acid chemoreception: a potential factor in the development of obesity? *Nutrients* 2013, 5: 1287-1300.
[41] Iqbal, J. & Hussain, M.M. Intestinal lipid absorption. *Am. J. Physiol. Endocrinol. Metab.* 2009, 296: E1183-E1194.
[42] Mizushige, T., Inoue, K. & Fushiki, T. Why is fat so tasty? Chemical reception of fatty acid on the tongue. *J. Nutr. Sci. Vitaminol.* (Tokyo). 2007, 53: 1–4.
[43] Manabe, Y., Matsumura, S. & Fushiki, T. Preference for high-fat food in animals. In: Montmayeur, J.-P. & le Coutre J., Editors. Fat detection: Taste, texture, and post ingestive effects. Boca Raton, FL. CRC Press. 2010, pp. 243-264.
[44] Gilbertson, T.A., Liu, L., Kim, I., Burks, C.A. & Hansen, D.R. Fatty acid responses in taste cells from obesity-prone and -resistant rats. *Physiol. Behav.* 2005, 86: 681-690.
[45] Gilbertson, T.A. & Khan, N.A. Cell signaling mechanisms of oro-gustatory detection of dietary fat: advances and challenges. *Prog. Lipid Res.* 2014, 53: 82-92.
[46] Febbraio, M., Guy, E., Coburn, C., Knapp, F.F. Jr., Beets, A.L, Abumrad, N.A. & Silverstein, R.L. The impact of overexpression and deficiency of fatty acid translocase (FAT)/CD36. *Mol. Cell. Biochem.* 2002, 239: 193-197.
[47] Gaillard, D., Laugerette, F., Darcel, N., El-Yassimi, A., Passilly-Degrace, P., Hichami, A., Khan, N.A., Montmayeur, J.-P. & Besnard, P. The gustatory pathway is involved in CD36-mediated orosensory perception of long-chain fatty acids in the mouse. *FASEB J.* 2008, 22: 1458-1468.
[48] Su, X. & Abumrad N.A. Cellular fatty acid uptake: a pathway under construction. *Tr. Endocrinol. Metab.* 2009, 20: 72–77.
[49] Martin, C., Chevrot, M., Poirier, H., Passilly-Degrace P., Niot, I. & Besnard, P. CD36 as a lipid sensor. *Physiol. Behav.* 2011, 105: 36-42.

[50] Nergiz-Unal, R., Rademakers, T., Cosemans, J.M. & Heemskerk, J.W. CD36 as a multiple-ligand signaling receptor in atherothrombosis. *Cardiovasc. Hematol. Agents Med. Chem.* 2011, 9: 42-55.

[51] Sundaresan, S., Shahid, R., Riehl, T.E., Chandra, R., Nassir, F., Stenson, W.F., Liddle, R.A. & Abumrad, N.A. CD36-dependent signaling mediates fatty acid-induced gut release of secretin and cholecystokinin. *FASEB J.* 2013, 27: 1191-1202.

[52] Kuda, O., Pietka, T.A., Demianova, Z., Kudova, E., Cvacka, J., Kopecky, J.& Abumrad, N.A. Sulfo-*N*-succinimidyl oleate (SSO) inhibits fatty acid uptake and signaling for intracellular calcium via binding CD36 lysine 164: SSO also inhibits oxidized low density lipoprotein uptake by macrophages. *J. Biol. Chem.* 2013, 288: 15547-15555.

[53] Abdoul-Azize, S., Selvakumar, S., Sadou, H., Besnard, P. & Khan, N.A. Ca^{2+} signaling in taste bud cells and spontaneous preference for fat: unresolved roles of CD36 and GPR120. *Biochimie* 2014, 96: 8-13.

[54] Nergiz-Unal, R., Lamers, M.M., Van Kruchten, R., Luiken, J.J., Cosemans, J.M., Glatz, J.F., Kuijpers, M.J. & Heemskerk, J.W. Signaling role of CD36 in platelet activation and thrombus formation on immobilized thrombospondin or oxidized low-density lipoprotein. *J. Thromb. Haemost.* 2011, 9: 1835-1846.

[55] Fukuwatari, T., Kawada, T., Tsuruta, M., Hiraoka, T., Iwanaga, T., Sugimoto, E. & Fushiki, T. Expression of the putative membrane fatty acid transporter (FAT) in taste buds of the circumvallate papillae in rats. *FEBS Lett.* 1997, 414: 461-464.

[56] Lobo, M.V., Huerta, L., Ruiz-Velasco, N., Teixeiro, E., de la Cueva, P., Celdrán, A., Martín-Hidalgo, A., Vega, M.A. & Bragado, R. Localization of the lipid receptors CD36 and CLA-1/SR-BI in the human gastrointestinal tract: towards the identification of receptors mediating the intestinal absorption of dietary lipids. *J. Histochem. Cytochem.* 2001, 49: 1253-1260.

[57] Chen, C.S., Bench, E.M., Allerton, T.D., Schreiber, A.L., Arceneaux, K.P. 3rd & Primeaux, S.D. Preference for linoleic acid in obesity-prone and obesity-resistant rats is attenuated by the reduction of CD36 on the tongue. *Am. J. Physiol. Regul. Integr. Comp. Physiol.* 2013, 305: R1346-R1355.

[58] Keller, K.L, Liang, L.C., Sakimura, J., May, D., van Belle, C., Breen, C., Driggin, E., Tepper, B.J., Lanzano, P.C., Deng, L. & Chung, W.K. Common variants in the CD36 gene are associated with oral fat

perception, fat preferences, and obesity in African Americans. *Obesity* (Silver Spring). 2012, 20: 1066-1073.

[59] Itoh, Y., Kawamata, Y., Harada, M., Kobayashi, M, Fujii, R., Fukusumi, S., Ogi, K., Hosoya, M., Tanaka, Y., Uejima, H., Tanaka, H., Maruyama, M., Satoh, R, Okubo, S., Kizawa, H., Komatsu, H., Matsumura, F., Noguchi, Y., Shinohara, T., Hinuma, S., Fujisawa, Y. & Fujino, M. Free fatty acids regulate insulin secretion from pancreatic beta cells through GPR40. *Nature* 2003, 422: 173-176.

[60] Cartoni, C., Yasumatsu, K., Ohkuri, T., Shigemura, N., Yoshida, R., Godinot, N., le Coutre, J., Ninomiya, Y. & Damak, S. Taste preference for fatty acids is mediated by GPR40 and GPR120. *J. Neurosci.* 2010, 30: 8376-8382.

[61] Sclafani, A., Zukerman, S., Glendinning, J.I. & Margolskee, R.F. Fat and carbohydrate preferences in mice: the contribution of alpha-gustducin and Trpm5 taste-signaling proteins. *Am. J. Physiol. Regul. Integr. Comp. Physiol.* 2007, 293: R1504-R1513.

[62] Liu, P., Shah, B.P., Croasdell, S. & Gilbertson, T.A. Transient receptor potential channel type M5 is essential for fat taste. *J. Neurosci.* 2011, 31: 8634-8642.

[63] Hamilton, J.A. & Kamp, F. How are free fatty acids transported in membranes? Is it by proteins or by free diffusion through the lipids? *Diabetes* 1999, 48: 2255-2269.

[64] Kamp, F. & Hamilton, J.A. How fatty acids of different chain length enter and leave cells by free diffusion. *Prostaglandins Leukot. Essent. Fatty Acids* 2006, 75: 149-159.

[65] Lieckfeldt, R., Villalaín, J., Gómez-Fernández J.C. & Lee, G. Apparent pKa of the fatty acids within ordered mixtures of model human stratum corneum lipids. *Pharm. Res.* 1995, 12: 1614-1617.

[66] Green, B.G. & Schullery, M.T. Stimulation of bitterness by capsaicin and menthol: differences between lingual areas innervated by the glossopharyngeal and chorda tympani nerves. *Chem. Senses* 2003, 28: 45-55.

[67] Bradley, R.M. & Grabauskas, G. Neural circuits for taste. Excitation, inhibition, and synaptic plasticity in the rostral gustatory zone of the nucleus of the solitary tract. *Ann. NY Acad. Sci.* 1998, 855: 467-474.

[68] Yarmolinsky, D.A., Zuker, C.S. & Ryba, N.J.P. Common sense about taste: From mammals to insects. *Cell* 2009, 139: 234-244.

[69] Johnson, P.M. & Kenny, P.J. Dopamine D2 receptors in addiction-like reward dysfunction and compulsive eating in obese rats. *Nat. Neurosci.* 2010, 13: 635-641.

[70] Laugerette, F., Passilly-Degrace, P., Patris, B., Niot, I., Febbraio, M., Montmayeur, J.-P. & Besnard, P. CD36 involvement in orosensory detection of dietary lipids, spontaneous fat preference, and digestive secretions. *J. Clin. Invest.* 2005, 115: 3177–3184.

[71] Lawton, C.L., Delargy, H.J., Brockman, J., Smith, F.C. & Blundell, J.E. The degree of saturation of fatty acids influences post-ingestive satiety. *Br. J. Nutr.* 2000, 83: 473-482.

[72] Ackroff, K., Vigorito, M. & Sclafani, A. Fat appetite in rats: the response of infant and adult rats to nutritive and non-nutritive oil emulsions. *Appetite* 1990, 15: 171–188.

[73] Magne, F.C., Mod, R.R. & Skau, E.L. Some *N*-disubstituted amides of long-chain fatty acids as vinyl plasticizers. *Indust. Eng. Chem.* 1958, 50: 617-618.

[74] Gerhold, K.A. & Bautista, D.M. Molecular and cellular mechanisms of trigeminal chemosensation. *Ann. N.Y. Acad. Sci.* 2009, 1170: 184-189.

[75] Wajid, N.A. & Halpern, B.P. Oral cavity discrimination of vapor-phase long-chain 18-carbon fatty acids. *Chem. Senses* 2012, 37: 595-602.

[76] Bacon, A.W., Miles, J.S. & Schiffman, S.S. Effect of race on perception of fat alone and in combination with sugar. *Physiol. Behav.* 1994, 55: 603-606.

[77] Running, C.A., Mattes, R.D. & Tucker, R.M. Fat taste in humans: sources of within- and between-subject variability. *Prog. Lipid Res.* 2013, 52: 438-445.

[78] Mattes, R.D. Oral thresholds and suprathreshold intensity ratings for free fatty acids on 3 tongue sites in humans: implications for transduction mechanisms. *Chem. Senses* 2009, 34: 415-423.

[79] Motlekar, N.A. & Youan, B.B. The quest for non-invasive delivery of bioactive macromolecules: a focus on heparins. *J. Control Release* 2006, 113: 91-101.

[80] Shenkin, A. & Wretlind, A. Parental nutrition. *World Rev. Nutr. Diet* 1978, 28: 1-111.

[81] Mathiowitz, E., Jacob, J.S., Jong, Y.S., Carino, G.P., Chickering, D.E., Chaturvedi, P., Santos, C.A., Vijayaraghavan, K., Montgomery, S., Bassett, M. & Morrell, C. Biologically erodable microspheres as potential oral drug delivery systems. *Nature* 1997, 386: 410–414.

[82] Fini, A., Cavallari, C., Rabasco Alvarez, A.M. & Rodriguez, M.G. Diclofenac salts, part 6: Release from lipid microspheres. *J. Pharm. Sci.* 2011, 100: 3482-3494.

[83] Lokamatha Swamy, K.M., Satyanath, B., Shantakumar, S.M., Manjula, D., Mohammedi, H. & Farhana, A. Matrix embedded microspherules containing indomethacin as controlled drug delivery systems. *Cur. Drug Deliv.* 2008, 5: 248-255.

[84] Waters, L.J. & Pavlakis, E. *In vitro* controlled drug release from loaded microspheres-dose regulation through formulation. *J. Pharm. Pharm. Sci.* 2007, 10: 464-472.

[85] Robson, H., Craig, D.Q.M. & Deutsch, D. An investigation into the release of cefuroxime axetil from taste-masked stearic acid microspheres. II. The effects of buffer composition on drug release. *Int. J. Pharm.* 2000, 195: 137-145.

[86] Robson, H.J., Craig, D.Q.M. & Deutsch, D. An investigation into the release of cefuroxime axetil from taste-masked stearic acid microspheres. Part 1: The influence of the dissolution medium on the drug release profile and the physical integrity of the microspheres. *Int. J. Pharm.* 1999, 190: 183–192.

[87] Robson, H., Craig, D.Q.M. & Deutsch, D. An investigation into the release of cefuroxime axetil from taste-masked stearic acid microspheres. III: The use of DSC and HSDSC as means of characterizing the interaction of the microspheres with buffered media. *Int. J. Pharm.* 2000, 201: 211-219.

[88] Powell, D.A., Nahata, M.C., Powell, N. E. & Ossi, M.J. The safety, efficacy, and tolerability of cefuroxime axetil suspension in infants and children receiving previous intravenous antibiotic therapy. *Ann. Pharmacother. (DICP)* 1991, 25: 1236–1238.

[89] Gooch, W.M. 3rd, McLinn, S.E., Aronovitz, G.H., Pichichero, M.E., Kumar, A., Kaplan, E.L. & Ossi, M.J. Efficacy of cefuroxime axetil suspension compared with that of penicillin V suspension in children with group A streptococcal pharyngitis. *Antimicrob. Agents Chemother.* 1993, 37: 159–163.

[90] Shalit, I., Dagan, R., Engelhard, D., Ephros, M. & Cunningham, K. Cefuroxime efficacy in pneumonia: sequential short-course i.v./oral suspension therapy. *Isr. J. Med. Sci.* 1994, 30: 684–689.

[91] Angadi, S.C., Manjeshwar, L.S. & Aminabhavi, T.M. Stearic acid-coated chitosan-based interpenetrating polymer network microspheres:

controlled release characteristics. *Ind. Eng. Chem. Res.* 2011, 50: 4504–4514.

[92] Kamphuis, M.M., Saris, W.H. & Westerterp-Plantenga, M.S. The effect of addition of linoleic acid on food intake regulation in linoleic acid tasters and linoleic acid non-tasters. *Br. J. Nutr.* 2003, 90: 199-206.

[93] Reed, D.R. Heritable variation in fat preference. In: Montmayeur, J.-P. & le Coutre, J., Editors. Fat detection: Taste, texture, and post ingestive effects. Boca Raton, FL. CRC Press. 2010, pp. 395–415.

[94] Drewnowski, A., Halmi, K.A., Pierce, B., Gibbs, J. & Smith, G.P. Taste and eating disorders. *Am. J. Clin. Nutr.* 1987, 46: 442-450.

[95] Drewnowski, A. Sensory preferences for fat and sugar in adolescence and adult life. *Ann. N.Y. Acad. Sci.* 1989, 561: 243-250.

[96] Frank, M.E. & Hettinger, T.P. What the tongue tells the brain about taste. *Chem. Senses* 2005, 30 (Suppl. 1): i68-i69.

In: Stearic Acid ISBN: 978-1-63463-172-3
Editors: Yunfeng Lin and Qiang Peng

Chapter 2

STEARIC ACID IN RUBBER CHEMISTRY AND TECHNOLOGY

E. Djagarova[1], D. Zheleva[1] and N. Tipova[2]
[1]University for Chemical Technology and Metallurgy, Sofia, Bulgaria
[2]University for National and World Economy, Sofia, Bulgaria

ABSTRACT

Recent development in properties research of stearic acid (SA) and its influence on the chemistry and technology of rubber compounds is reviewed. The stearic acid is involved in almost all rubber compounds. It is used in relatively low amounts (up to 3.0 phr) but has a multifunctional effect on their properties and processing. The function of SA as a softener and a filler dispersing agent within rubber compounds is examined. As a softener, SA influences the viscosity of filled compound. Processing is facilitated because of its decrease during mixing. Such influence of SA on rheological properties of rubber compounds is manifested in some uncommon results of the apparent viscosity, determined at low shear rate values. These results could be explained when examined in common with the experimental results for the dispersing influence of SA onto fillers and the amount of bond rubber. Based on the rheological properties of rubber compounds in dependence of the SA amounts within, a method is proposed for determining the optimal SA amount as a dispersing agent.

Contemporary science considers SA an effective activator (along with ZnO) of sulfur vulcanization of non-saturated rubbers. With regards to the mechanism accelerated sulfur vulcanization, most of the presently proposed hypotheses suggest that at the conditions of the process a reaction occurs between SA and ZnO which results in forming zinc stearate. SA salt probably participates in the formation of an activating complex with the appropriated accelerator and sulfur which provokes the crosslinking of rubber macromolecules. Research conducted on rubber vulcanization with zinc stearate as an activator within the compound (but without ZnO) shows that vulcanizates of good properties are obtained. Using zinc stearate instead of ZnO for rubber vulcanization could cause a very serious environmental problem. In this case, the amount of zinc within the rubber compounds is reduced by more than 10 times. This will consistently decrease environment pollution with toxic zinc ions which results from tire tread wear during traffic. The SA salt could resolve one more environmental problem. Parts of waste tires are recycled and one of the obtained products is rubber flour. A method for modification of its surface by means of powdering the particles with zinc stearate is elaborated on and the use of rubber flour in amounts up to 20 phr becomes possible.

INTRODUCTION

Modern civilization is impossible without rubber products, such as tires, conveyor belts, hoses, seals. Each of these products is produced from different rubber compounds or blends of variable composition. The formulation of each rubber compound includes a lot of different ingredients. The tire tread could include about 15–18 different ingredients and the stearic acid is constantly present within the formulation. Its amount within the rubber compounds is not high. It is used in amounts up to 3 phr but shows a multifunctional effect [1 - 5]. By 1920, black-reinforced tread compounds were in general usage and a fast development of the rubber industry begins. Stearic acid starts to function as filler dispersing agent within the rubber and as a softener for filled rubber compounds. Subsequent research on the mechanism of sulfur accelerated vulcanization of non-saturated rubbers shows that stearic acid (along with ZnO) has an important role as an activator of the vulcanization. Sulfur vulcanization was already discovered in 1839 in USA by Charles Goodyear. There have been a lot of hypotheses proposed for the vulcanization mechanism, but the complex of chemical reactions occurring within a heterogeneous and multicomponent system such as the rubber compound has not been explained until now. However, the opinion of researchers on the key role of stearic acid as an activator of the sulfur accelerated vulcanization has become more and more accepted. Most of the presently proposed hypotheses, with regards to mechanism of accelerated sulfur vulcanization, suggest that at the conditions of the process a reaction occurs between stearic acid and ZnO and zinc stearate is obtained. The stearic acid salt probably participates in the forming of an activating complex with the appropriated accelerator and sulfur which provokes the crosslinking of the rubber macromolecules.

Acting as a softener, stearic acid increases rubber plasticity and the capacity of the ingredients to disperse within the rubber. It decreases the softening temperature of the rubber compound at the initial vulcanization period and the level of contracting of rubber compounds when molded. It decreases the heat build-up during the process of mixing, the time necessary for preparing the compounds and the energy consumption. Stearic acid not only disperses the ingredients within the compound but also wets the filler's particles and thus reinforces their bonding with the rubber's macromolecules. The stearic acid, similarly to other softeners, does not influence glass transition temperature of the rubbers. Therefore, it does not improve their cold-resistance.

It is typical for the rubber industry that the important inventions are manifested firstly in technologies with a practical aspect and only after that the research on their theoretical explanation gets developed. It is clear for each specialist from the field of rubber products that the content of the stearic acid and/or zinc stearate within the rubber compound contributes to easily release the article from the mold and the final product has a better appearance. With changes in the amounts of stearic acid or of zinc stearate a decrease of the pollution of the mold working surface is achieved which has a substantial economic effect.

Taking into account the statements above it is to be concluded that the stearic acid is an ingredient for rubber compounds or blends of multifunctional effect. The stearic acid is:

- A dispersing agent for the fillers within rubber compounds;
- A softener of filled rubber compounds;
- An activator of sulfur accelerated vulcanization of non-saturated rubbers;
- Release mold agent in the production.

On a yearly base, the world consumption of stearic acid for rubber articles production is more than 200,000 tons. These are tentative figures, given that each rubber compound contains 1.5 phr of stearic acid and the processed non-saturated rubbers are about 15 million tons per year.

Stearic Acid – Softener and Dispergator

The wide spread use of synthetic rubbers became possible with the use of reinforcing fillers. The purpose of the fillers within rubber compounds is manifested in two ways:

1. A substantial increase of the mechanical properties of the vilcanizates is observed which results from including fillers within the rubber compounds.
2. The fillers decrease the price of rubber products because it is many times lower than that of rubber itself.

The use of carbon black as a reinforcing filler for tire treads started back in 1918. The effect of the reinforcing by carbon black is not clear enough. It is known that the factors influencing the reinforcement are as follows:

1. Filler's particles size;
2. Filler's structure;
3. Interaction of the filler's particles with the rubber macromolecules.

The fillers are characterized by primary and secondary structure. The primary structure of carbon black is manifested by chains comprised of individual particles. This structure cannot be destroyed under mechanical influence. The secondary structure represents bigger or smaller agglomerates which are destroyed under the dispersing effect of mixing the rubber with the filler. The mixing is facilitated by using different substances such as dispersing agents. Among all the substances tested until now stearic acid is used the most.

At the same time the stearic acid acts as a softener. It became known that softeners within the rubber compound decrease the duration of preparing the compounds as well as the energy consumption. Fat acids (especially the stearic acid) wet the particles of the filler and reinforce their bonds with the rubber. There are different ways of determining the dispersion within the rubber of powder fillers. Most frequently, measuring the mechanical properties of the vulcanizates is applied as an indirect method for the regular particles distribution and depending on the duration of the mixing process. Other methods such as atomic force microscopy of carbon black aggregates [6] are elaborated on as well. Research has been conducted [7] on the effectiveness of 4 methods from the point of view of reproducibility, mathematical expression and flow rate for compounds of NR (natural rubber), EPDM (ethylene propylene triple co-polymer), NBR (acrylonitrile butadiene rubber) and SBR (styrene butadiene rubber)/BR (butadiene rubber)/NR and carbon black depending on the duration of the mixing. For this purpose commonly used are the methods of measuring the electric conductivity or the electrical resistance of carbon black filled vulcanizates based on NR, EPDM, NBR and SBR [6-9]. It became known that single polymers are isolators of specific volume of electrical resistance $10^{15} - 10^{18}$ Ω. Consequently, the electric conductivity of carbon black filled rubbers is due especially to the filler particles. The results for the electrical resistance depend on the type and amount of carbon black within the compounds. It was stated [8, 9] that the curves of the mechanical properties and of the electric conductivity have one and the same characteristic depending on the duration of the mixing. It is suggested that measuring the electric conductivity in this case is a fast method for determining the carbon black dispersion within the rubber and the maximum electric conductivity attained could be a criterion for optimal duration of processing the corresponding mixer.

The effect of curing additives on the dispersion kinetics of carbon black in styrene butadiene rubber (SBR) compounds was investigated by means of the method of the online measured electrical conductance [10]. Addition of curing additives such as stearic acid and diphenylguanidine (DPG) accelerates the carbon black dispersion process significantly. The viscosity of the rubber matrix was not changed after their addition. The addition of stearic acid and DPG may alter the filler–filler interaction that consequently leads to faster dispersion processes. The obtained difference in morphologies of SBR mixtures containing stearic acid and DPG respectively are caused by their different infiltration behavior which may lead to different dispersion mechanisms. Addition of ZnO could not improve the dispersion process of carbon black because of its limited interaction with carbon black.

Most current methods used to measure the degree of dispersion are based on some form of microscopy – light microscopy (at small magnification) of cryomicrotomed rubber specimens in transmitted light [11]. According to authors [12] capillary rheometry can be used to demonstrate differences between products acting as external and as internal lubricants in relation to viscosity and to extrudate swell.

The dispersion of carbon black agglomerates which are in suspension of liquid polymethylsiloxane and their dependence from the shear stress, has been studied by simple shear flow in cone-and-plate shearing device [13, 14]. The statements presented above show that the results of the research on the influence of mixing duration on the dispersing degree of the carbon black within the rubber have been published. Unfortunately, there have been no publications on the influence of the stearic acid amounts on the dispersing or on the viscosity of filled rubbers. One reason for that is probably the commonly adopted for industrial purposes amount of the stearic acid in rubber compounds which is up to 3 phr. Another reason is the probable opinion of the professional that such a small amount could not affect the compound's viscosity. Moreover, that aromatic and aliphatic oils in big amounts ranged about 10 % [4, 5] are used as softeners in highly filled compounds. In some older publications [15-18] there are results from investigations about the influence of small levels (up to 1%) of low molecular organic substances and powdered additives on the polymer melt viscosity. In such a way are established dependences with extreme points as a function of the additive's amounts. It was perceived later on that such effect of the additives could be observed for the viscosity of rubber compounds as well [19-22]. It is necessary that the additive's levels for rubbers be higher if compared to that for plastics, i.e., more than 1% because of the following:

1. The viscosity of the rubber compounds is several times higher than that of the polymer melts.
2. Part of the additives is adsorbed on the particles surface of the filler and could not affect the viscosity.

This was the ground on which were implemented investigations about the effect of the amount of stearic acid on the viscosity of carbon black filled compounds based on different rubbers [23- 26]. It is important to determine its optimal dosage when used as a dispersing agent taking into account its bi-functional behavior in the compounds.

Usually, the optimal amount of dispersing agent in a definite rubber compound is dependent from the type of the rubber used, from the type and the amount of the dispersing agent used. A dispersing agent dosage which is different from the optimal one could increase the energy consumption during the mixing of the rubber with the filler and thus affect the final product quality. When analyzing the dispersing capability of the stearic acid in carbon black filled rubber compounds it was stated that with increasing the amount of stearic acid the compound viscosity smoothly increases, it reaches maximum values and then smoothly decreases [23]. Taking into account these results and some information from the industry, a method for determining the optimal amount of dispersing agent in rubber compounds through evaluating the rheological properties of the compound and their dependence from the quantity of that dispersing agent [27] was elaborated on.

In an attempt to follow the effect of concentration of stearic acid on the rheological properties, compounds from SBR rubber -filled with 50 phr carbon black N550, ZnO, and with varying the stearic acid concentration (0; 0.5; 1.0; 1.5 and 2 phr) were prepared. Using Brabender plasticorder and capillary viscometer these compounds were studied. The observed effective (apparent) viscosity h_e (30÷120%) (Figure 1) and torque maxima M_b (10÷20%) (Figure 2) at 1.5 phr stearic acid were assigned to the dispersion effect of stearic acid during the process of mixing of carbon black with rubber.

This is confirmed by the significant dispersion of experimental data during determination of rheological properties of the compound of carbon black and rubber in absence of stearic acid, e.g., "zero compound" (Figure 3)

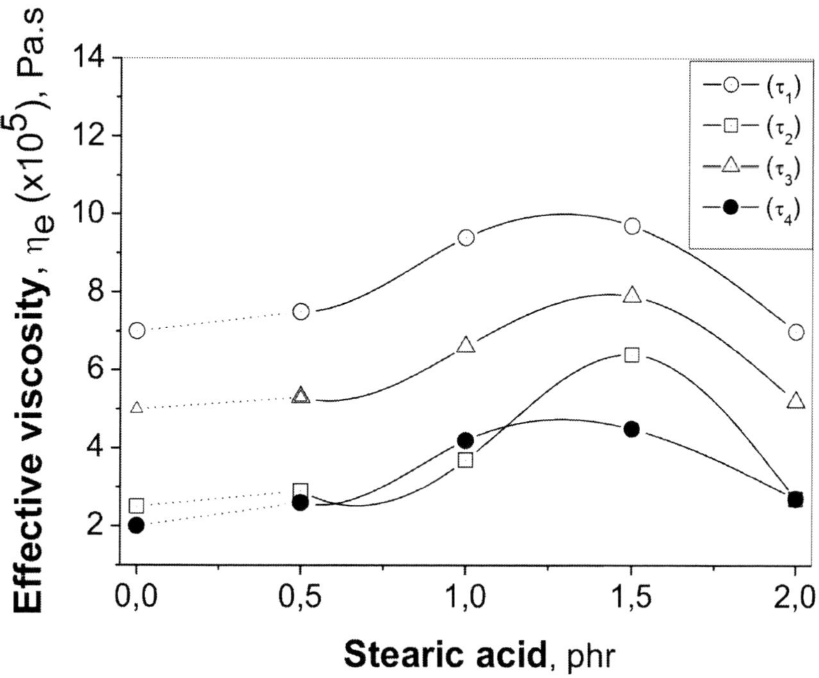

Figure 1. Dependence of effective viscosity η_e from stearic acid concentration in filled rubber compounds at different shear stresses (τ_1=1.4.10^5; τ_2=2.0.10^5; τ_3=2.0.10^5; τ_4=2.4.10^5; τ_5=3.0.10^5 Pa) and temperature 110 °C in capillary viscometer.

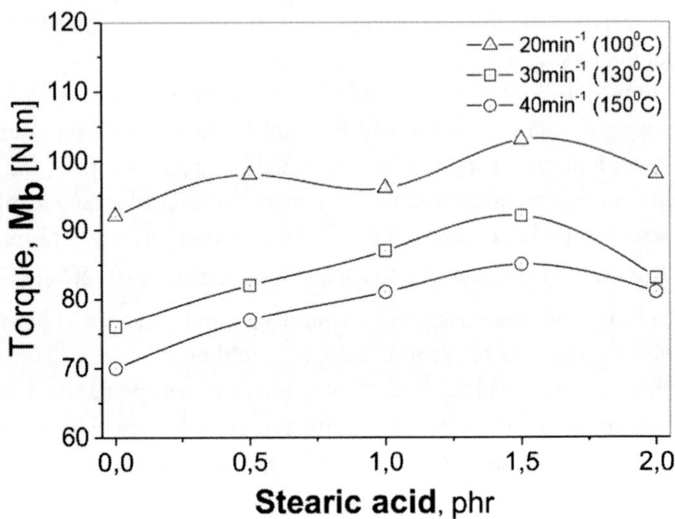

Figure 2. Dependence of torque M_b from stearic acid concentration in compounds at revolutions of 20, 30 and 40 min^{-1} in Plasticorder Brabender.

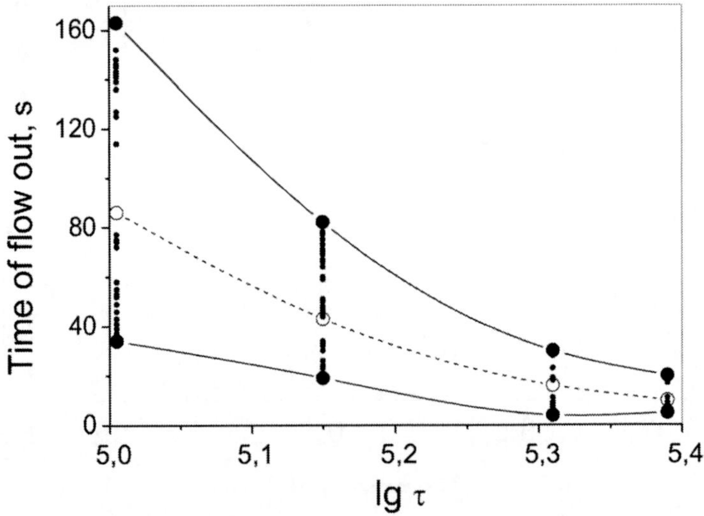

Figure 3. Experimental results from measuring the flow time of the "zero" rubber compound under different shear stresses (τ_1=1.4.10^5; τ_2=2.0.10^5; τ_3=2.0.10^5; τ_4=2.4.10^5; τ_5=3.0.10^5 Pa) and temperature 110 ° C.

The evaluation of the dispersing capability of some substances in regards to powder ingredients, according to some previous works [7-9], could be based on experimental results obtained during the determination of the specific volume electrical resistance ρ_V and the specific surface electrical resistance ρ_S. These authors have stated that with increasing the dispersion of the carbon black in rubber compounds the specific volume electrical resistance increases too.

The evaluation of the amount of bond rubber and determination of specific volume electrical resistance ρ_v (Figure 4) completely correlates with M_b and η_e dependences from stearic acid concentration. Based on all results obtained, it could be suggested that the optimal dispersing effect of stearic acid for this rubber SBR and for carbon black N550 is between 1.0 and 1.5 phr.

The increase of η_e for rubber compounds with increasing the stearic acid amount could be explained by a deep analysis of mixing phenomena of the ingredients. In general, mixing is the process of stirring the two or more components. Moreover, depending on the stirring conditions (temperature, stirring speed and viscosity of the components) a mechanical mixture with an adequate degree of homogeneity is given. The mixing of rubber with the ingredients could not be interpreted by traditional hydrodynamic analysis procedure because the process could not be considered as based on laminar or stationary ordered flow of viscose medium studied on the basis of classic hydrodynamic and rheology.

Two almost simultaneous processes could be distinguished during the preparation of rubber compounds– mixing and dispersion. The simple mixing is a process of steady distributed particles in the compound without decrease of initial particles size. Dispersion mixing is a process which leads to decrease of particles size and an increase of interfacial surface and some increase of compound homogeneity. The mixing of carbon black with rubber could be considered an example of this type of mixing.

The observed increase of η_e at 1.5 phr stearic acid in filled SBR blends could be explained with the increase of its dispersion effect. This means that the increase of concentration of stearic acid gives rise to degradation of agglomerates of carbon black which transform into aggregates with larger interfacial surface, a closer contact and stronger interaction with the macromolecules of rubber, respectively. This effect of the concentration of stearic acid is considered as reinforcing of the carbon black.

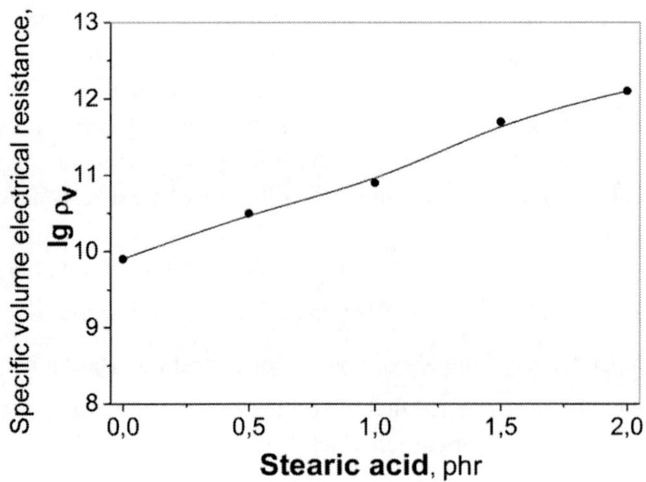

Figure 4. Dependence of specific volume electrical resistance $\lg \rho_v$ from stearic acid concentration in vulcanized compounds based on SBR.

The decrease (30-50 %) of η_e (Figure 1) of the compound at 2.0 phr stearic acid could be explained by the effect of the excess amount of stearic acid (over 1.5 phr) which acts as a softener. Increasing the mechanical values of vulcanizates containing up to 1.5 phr stearic acid confirmed the results of rheological tests [23]

It was established [28] that the dispersion action of stearic acid depends from the type of the carbon black but it does not depend from the type of elastomer.

In order to describe rubber compounds from the point of view of their processing capability different methods are used [29]. In this respect full characterization is achieved by analyzing the flow curves of rubbers and rubber compounds. The flow curves can be obtained by different viscometers. In the rubber industry (over 100 years) a technological index has been widely used – the Mooney viscosity. It serves for a comparison and control. It is measured relatively quickly with the respective viscometer and in accordance with the International standard ISO 289-1:2002. The obtained Mooney viscosity gives information just for one point of the flow curve of the rubber or rubber compound and this is the reason that no full rheological characteristics of the elastomeric material are obtained [30]. In order to describe the processing capability of rubber compounds by means of Brabender Plasticorder, the measurements are usually performed at a relatively high rotation speed – at about 20 min^{-1}. When these revolutions are retained the shear conditions during the production process are mostly similar to the processing conditions but are very different from the shear conditions during the measurement of the Mooney viscosity.

The previous investigations were carried out on a Plasticorder [24] at low rotor revolutions of 1, 2, 5 and 10 min^{-1}. This investigation included an attempt to calculate the average shear rate $\dot{\gamma}$ based on the experimentally obtained data at different rotor revolutions. It was found through extrapolation of the linear relationship shear rate $\dot{\gamma}$ v/s rotor speed that $\dot{\gamma}$ values in the range 1 ÷ 2 s^{-1} correspond to the rotor revolutions 1 to 5 min^{-1}. Taking into account that $\dot{\gamma}$ in the Mooney viscometer is at about 1-1.5 s^{-1}, we tried to find out a correlation between the values of the Mooney viscosity and the results from the measurements carried out on the Plasticorder.

In [31] a correlation is found between the Mooney viscosity and the apparent viscosity, obtained by means of a capillary viscometer. The investigations were carried out on both rheometers with 6 types of rubbers, namely NR crepe sheets (natural rubber), SMR-20 (Standard Malaysian Rubber, i.e., type natural rubber), SBR (styrene-butadiene rubber), NBR (acrylonitrile-butadiene rubber), MQ (silicone rubber), IIR 268 R (butyl rubber) and their compounds. The following equation was obtained:

$$PL = 10(\lg \eta + t)$$

where

PL – Mooney viscosity, Mooney units;

η – Apparent viscosity, Pa.s;

t – Time for flow out of material in capillary viscometer, s.

According to the authors [31], the error does not exceed 2%.

Other researchers [32] found a correlation between rheological behavior of filled rubber blends and their properties/structure. They suggested that a thorough understanding of the characteristic rheological response to the morphology/structure evolution of multiphase/multi-component polymers facilitates researchers optimizing the morphology/structure and ultimate mechanical properties of polymer materials.

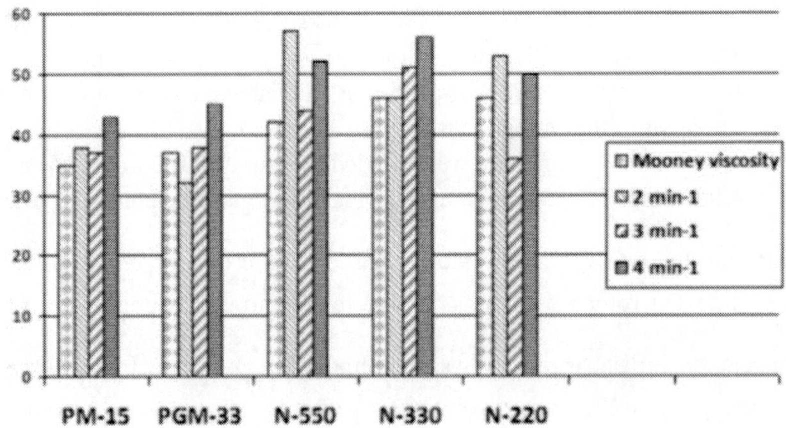

Figure 5. Values of Mooney viscosity ML and torque MB for the SBR compounds filled with 50 phr of different types of carbon black (PM-15, PGM-33, N-550, N-330 and N-220) at different rotors revolutions (2, 3, 4 min-1).

Similar investigations aimed at finding a correlation between the Mooney viscosity and the torque were carried out for NR (natural rubber) and BR (butadiene rubber) compounds on a Plasticorder in the monograph [33]. According to the author, the deformation conditions strongly influence the compounds structure, respectively their effective viscosity. This is the reason to find just qualitative dependences property/composition.

A correlation was discovered [34] between the Mooney viscosity ML and the torque M_B on a Plasticorder and the corresponding equations for this dependency were derived for rubber compounds based on SBR (styrene-butadiene rubber) containing different types of carbon black and for compounds with one and the same type of carbon black but of different level of filling.

A correlation was revealed between ML and M_B for SBR rubber compounds with N-550 carbon black at the levels of 30, 40, 50, 60 and 70 phr and an equation for this dependency was developed:

$$ML = 1.8\, M_B - 45.2 \text{ [34].}$$

Here presented phenomena and objectives are especially related to carbon black as filler. In the rubber industry different fillers are also usually used basically mineral fillers but in considerably lower amounts. It came to be known that if bright fillers are used for rubber compounds filling [5], the following peculiarities should be taken into account:

- The viscosity of the compounds for several fillers increases much more if compared with carbon black;
- They disperse within the rubber with more difficulty;
- They often slow up the vulcanization.

By reason of the adsorbed humidity the active bright fillers also contain crystallized water. If dehydration is applied after that the dispersion becomes more difficult. That is why, when processing such rubber compounds there are a lot of peculiarities observed:

- The compound of the rubber and the filler (with no other ingredients) is prepared firstly and it remains for about 24 hours.
- The softeners, plasticizers and others are added after that. If these ingredients are added together with the filler, they are adsorbed on its surface and thus screen the access of the rubber to the filler. The result of this is that there is an obstacle for the creation of the bonds rubber-filler.
- The precipitated silica and ZnO interact with each other and this decreases the reinforcement effect of the filler.
- By reason of the adsorption of the stearic acid and the accelerators of the vulcanization on the part of the mineral active fillers, these ingredients are used in higher levels.

By the end of the last century intensive investigations on the precipitated silica as filler for rubber compounds were launched. This active development of the investigations was due essentially to the possibility to replace part of the carbon black by silica within tire treads. In result, the tire's rolling resistance decreased and this led to decreasing the fuel consumption by approximately 5%, decreasing the amount of the damaging environmental gases and is of great ecological and economic importance. That is why the tires produced were called "green tires". The decrease of the tire's rolling resistance became possible with the use of bi-functional organic silanes. There are OH groups on the silica surface which lead to bonds creation between the filler's particles. This significantly exposes their dispersion within the rubber. The poor filler dispersion leads to decrease of its reinforcing effect and especially to increase of the heat build-up of the rubber vulcanizates under dynamic loads. For achieving a better dispersion of the filler within the rubber a plurality of low molecular substances, usually organic silanes, are used. There is a concept that during the mixing process the organic silanes react only with the surface part of the filler but the reaction with the polymer itself occurs during the vulcanization. This is how the silanes create a "bridge" between the filler's particles and the rubber macromolecules which results in reinforcing effect. There are some references to the fact those additives such as the stearic acid [35], ZnO [36, 37], CBS [38, 39] modify the surface stress of silica [40]. The effect of different additives on the selective distribution of the silica within both phases of the blend of NBR and NR is investigated in dependence from the correlation additive/silica [41].

Stearic Acid – Activator of the Sulfur Vulcanization of Non-Saturated Rubbers

The vulcanization is a complex of processes and as a result of that the rubber, being plastic and a soluble mass in organic solvents, is transformed into a non-melting, insoluble and elastic material. These processes occur at high temperature and pressure with the participation of a vulcanizing agent, most frequently sulfur, accelerators, and activators. These changes are due to the building of –C-C-, -C-S-C-, -C-S-S-C- and –C-S_n-C- crosslinking bonds among the rubber macromolecules.

In 1839, Goodyear firstly ascertained that, when heated, the sulfur changes the rubber. Charles Goodyear is best known for one unique patent, U.S. patent 3633, "Improvements in India Rubber Fabrics Vulcanization of Rubber" (1849). [42]. There are different descriptions of this invention and at the very beginning even Goodyear called it "metallization" as an analogy to steel tempering [43]. Goodyear sent samples to an agent of his in Europe. Some of them got into the hands of Hancook who did his best to achieve the same result and finally managed to do so. As it was, he hurriedly preserved the priority right of the invention and granted it in 1843 by the name of "vulcanization". At the beginning, the vulcanization was considered an adsorption process. It is Weber who firstly developed the idea that basically the vulcanization is a chemical reaction between sulfur and rubber [44]. He explored in a wide rate investigation the nature of vulcanization itself. He implemented vulcanization of different compounds containing various amounts of sulfur, modifying the duration time and the temperature, thus determining the level of the sulfur bonded. Performing the graphics of the correlation between the percentages of the bonded sulfur and the empiric formulae of the resulting sulfide he got a saw-tooth curve which, according to him, corresponded to definite chemical compounds of the sulfur and the rubber. Weber's ideas were further developed by Kirhov, Mayer and Mark who stated that during the vulcanization sulfur is bonded with the rubber's double bonds thus forming sulfides. [2].

After discovering of the vulcanization the rubber industry watches it's time and it is only after the first decade of the 20 century that another 3 great inventions lead to improved features and stability of the rubber. These inventions are regarding: the organic accelerators, the reinforcing activity of carbon black as filler and the anti-ageing agents. About 1910 it was discovered that hot vulcanization is accelerated by a range of substances which not only accelerate the vulcanization rate but influence positively the properties of the resulting vulcanizates. It was found in 1920 that if ZnO and fat acid, especially stearic acid, are added the vulcanization process is strongly activated and the properties of the final product are highly improved. Thus, during the next 70 years the vulcanization process was developed and improved more and more. This is clearly demonstrated within the following table 1 [45].

For the past several decades' scientific research on the vulcanization mechanism was carried out but until now there has been no unified theory about it. Moreover, for some accelerators there are several different hypotheses regarding the mechanism of vulcanization of different rubbers and even different hypotheses for a particular accelerator with a particular rubber. It became known that the effect of the accelerators on vulcanization is multifunctional and depends on the type of rubber, the type of accelerator, the nature of the filler and vulcanization conditions, but the mechanism of reacting is complex and not sufficiently explained. It was discovered that the accelerators show their full activity in the presence of other substances which are the so called activators. For the past several decades' scientific research on the vulcanization mechanism was carried out but until now there has been no unified theory about it. Moreover, for some accelerators there are several different hypotheses regarding the mechanism of vulcanization of different rubbers and even different hypotheses for a particular accelerator with a particular rubber.

It became known that the effect of the activators on vulcanization is multifunctional and depends on the type of rubber, the type of accelerator, the nature of the filler and vulcanization conditions but the mechanism of reacting is complex and not sufficiently explained.

Table 1. Reduction of vulcanization time obtained over 70 years [45]

Year	1850	1880	1905	1920
Natural rubber	100	100	100	100
Sulfur	8	8	6	3
Zinc oxide	-	5	5	5
Aniline	-	-	2	-
Stearic acid	-	-	-	1
2-mercaptobenzothiazol (MBT)	-	-	-	1
Optimal vulcanization time at 142°C, min	360	300	180	30

It was also discovered that small amounts of ZnO (in combination with stearic acid) decrease the duration of vulcanization and improve the properties of the vulcanized products, even in case of non-accelerated vulcanization. Fat acids (stearic acid in particular) are used as an additive for a better dispersion of the ZnO within the system and help in forming a complex of free Zn ions with the accelerators [46]. Nowadays, it is known that a lot of metal oxides such as ZnO, MgO, CdO, CaO, PbO etc. have an activator effect. The essential representative from this group is the ZnO. Almost all the formulations for rubber articles contain this metal oxide. It has been known that the stearic acid is in dosage up to 3 phr. In the last few years the results which were received from researching the effect of the stearic acid on some properties of the compounds and the vulcanizates in amounts over 3 phr were published [47, 48].

Strain-induced crystallization of cross linked natural rubber (NR) and its synthetic analogue *cis*-1,4-polyisoprene (IR), both mixed with various amounts of stearic acid (0 – 4 phr) were investigated [47] by time resolved X-ray diffraction and simultaneous mechanical (tensile) measurement. No acceleration or retardation was observed on NR in spite of the increase of stearic acid amount. Even the stearic acid-free IR crystallized upon stretching and the overall crystallization behavior of IR shifted to the larger strain ratio with the increase of stearic acid content. These results suggest that the extended network chains are effective for the initiation of crystallization upon stretching, whereas the role of stearic acid is inconsequential. These behaviors are more different from their crystallization at low temperatures by standing where the stearic acid acts as a nucleating agent.

An investigation [48] was carried out comparing seven different natural rubber formulations. In each case stearic acid levels used varied from 0 to 6 phr. The results show that the stearic acid level of 4 phr gives vulcanisates optimal physical properties such as hardness, tensile strength, modulus, and abrasion resistance. But above this level and up to 6 phr abrasion resistance and compression set are also favored. However, they have adverse effect on those properties they favor at lower levels. Therefore, it can be concluded that the stearic acid level is to be based on the more desirable vulcanizate properties which could be specified by area of application.

The influence of stearic acid loading on the adhesion of rubber to brass-plated steel wires has been researched [70]. The so-called squalene method was employed to investigate the adhesion interface build-up during the vulcanization reaction. Variation of the stearic acid loading has a direct influence on the bonding interface and at the same time has a strong influence on the rubber properties. The surface of the sulfidated wires was analyzed using optical, focus variation and scanning electron microscopy coupled with energy dispersive X-ray spectroscopy. Increasing amounts of stearic acid accelerated the sulfidation reaction. Rubber properties and adhesion values were measured for natural rubber compounds with variable amounts of stearic acid. In most cases the adhesive strength exceeded the cohesive strength of the rubber.

It is known that zinc cations of the ZnO and/or zinc compounds react with the organic accelerators thus giving zinc complexes accelerators. This is one of the most important stages of the vulcanization scheme [50, 51]. It has been suggested that the better zinc dispersing in the system ends with zinc complex generating and the zinc ions are free to form active accelerator complexes. By reason of that, zinc ion is supposed to be the central atom in the accelerated sulfur vulcanization [46]. This idea has remained indivertible for the last 25 years.

Recently, more and more persistent attempts are made to find out how to replace the ZnO or to reduce its content in rubber compounds to a minimum. The main reason for it is that zinc is included in the list of substances which, without a doubt, effect negatively the environment [53]. Notwithstanding the fact that zinc is a heavy metal, the statement that it is dangerous for the environment is not correct. Practically zinc is a recyclable natural part of our environment. It is the 17-th most widespread element in the earth's crust and is an element important for people, animals and plants [54]. Illustration of this fact is that people who lack of zinc there have problems with growth, fertility, memory and immune system. As an element it is not toxic and only its ions are. Recently the Program on Chemical Safety (IPCS), established and especially designated for zinc, took a position which stated: "Zinc is an important and useful element of the environment. Possibilities for environmental excesses as well as insufficiencies are quite real. For this reason, the regulation of the environmental amounts of zinc is of great importance. We need to be protected against its toxic effect but at the same time it is necessary to control its quantity and to keep its level of the zinc above the prescribed critical minimum." [55]

The most striking example of the constant diffusion of zinc in the environment is the waste of tire treads. Professional data show that the weight of one car tire is about 7 kg. The tread of a car tire is about 30%. The calculations suggest that from each wasted tread approximately 2 kg vulcanizate are released in the environment which is equivalent to almost 50 g ZnO. Latest data [56] show that in the European Community about 300 000 tons of ZnO are used, and ¾ of this amount is used for the purposes of production of tires and for the rubber industry.

In series of works in recent years results of professional investigations on reduction of zinc amounts in the rubber compounds have been published. On this basis it was concluded that there are 3 directions:

- Reducing the ZnO amounts by using its derivative with a different specific surface area of the particles;
- Replacing the ZnO with other metal oxides;
- Investigations on different zinc compounds as activators of the vulcanization.

The paper [57] presents a comparison of 5 types of ZnO with different specific surface area. According to the results obtained for similar formulations it is possible to reduce the ZnO dosage down to 2 phr. This could be economically advantageous and could help solve some environmental problems as well.

Several research studies [58, 59 and 67] have been carried out in order to substitute it with different metal oxides such as MgO. The influence of different mixtures of ZnO and MgO on the vulcanization of natural rubber has been investigated and a model compound vulcanization has been used to study the role of MgO on the mechanism of vulcanization.

According to some authors, the ZnO and the stearic acid react with each other within the rubber compound during the vulcanization process and generate zinc stearate. The effect of zinc stearate as an activator for black filled SBR compounds with the sulphenamide accelerator TBBS has already been studied [52]. The authors of this work did not cite any other references on the topic. They examined the Zn stearate as an activator of the sulfur vulcanization in amounts from 10 to 20 phr. This choice is not peculiar and accounts for the expectation that 10 phr Zn stearate corresponds to the mole amount of Zn^{2+} ions compared to 1 phr ZnO. For such a concentration of the Zn stearate it has been stated that the vulcanization network density is lower than that formed when using ZnO. The influence of other zinc compounds as accelerators has been investigated as well [52]: zinc-m-glycerolate, zinc-2-ethylhexanoate, zinc borate and a new sulfur vulcanization activator [52], which is a clay mineral, charged with zinc ions on its surface, zinc monomethacrylate [66]

The influence of Zn stearate as an activator of sulfur vulcanization of filled styrene-butadiene rubber (SBR) accelerated with mercaptobenzothiazole (MBT), tetramethylthiuramdisulfide (TMTD) and sulphenamide accelerator (CBS) has been investigated [60-63]. On the basis of experimental data obtained by means of vulcameter during the investigation of vulcanization performances /ΔM = MH – ML/ for rubber compounds cured in the presence of an activator, it was discovered that their crosslink density for the amounts of 0.5, 1.0 and 2.0 phr Zn stearate is the same, compared to that of the sample and for compounds with 5 and 10 phr Zn stearate the density of the vulcanization network is higher. This conclusion was also confirmed by the results of molecular weight M_c of the segment between two cross-linking points obtained by means of the equilibrium swelling degree of the vulcanizates. The time t_{90} of the compounds containing Zn stearate is equal to the same time (from 10 to 13 minutes), the sample compound included. The cure rate V is highest for compounds with 2 phr Zn stearate and is almost the same as that of the sample compound (Figure 6).

Mechanical properties of the vulcanizates containing Zn stearate are close to the analogous characteristics of the sample compound. It is proved that Zn stearate is an effective dispersing agent for the carbon black in a wide range of concentrations within the rubber compound. Experimental results reveal that for the carbon black filled SBR compounds and the MBT Zn stearate is a suitable activator.

The concentration of the Zn for the range of the applied amounts of 0.5 phr to 10phr Zn stearate corresponds to ~0.06 phr and 1.25 phr ZnO. The use of Zn stearate does not lead to a risk for the vulcanization and for the mechanical performance of rubber products. If Zn stearate is used, it is possible to decrease the amount of the Zn from 4 to 20 times.

Application of Zn stearate, instead of ZnO and stearic acid, as an activator of sulfur MBT vulcanization reveals an additional range of advantages:

- Replacement of two ingredients (ZnO and stearic acid) with a single one (Zn stearate) is advantageous for the dosage and batching;
- The price of the compounds is lower because Zn stearate is cheaper than ZnO and stearic acid (because of their production technology);
- The Zn stearate in rubber compounds decreases the adhesion process in the mixing machines;
- Zn stearate makes easy the release of the vulcanizates from the molds when the vulcanization is already finished;
- Zn stearate decreases the mold surfaces fouling in result of the vulcanization, and increases repeatedly the time limit between two consecutive surface cleanings.

Zinc stearate was synthetized [64] by precipitation method through steps: neutralization of stearic acid by sodium hydroxide then double decomposition using zinc sulfate to precipitate zinc stearate. It was concluded that zinc stearate can be used as activator for sulfur vulcanization process instead of ZnO and stearic acid in absence and presence of fillers.

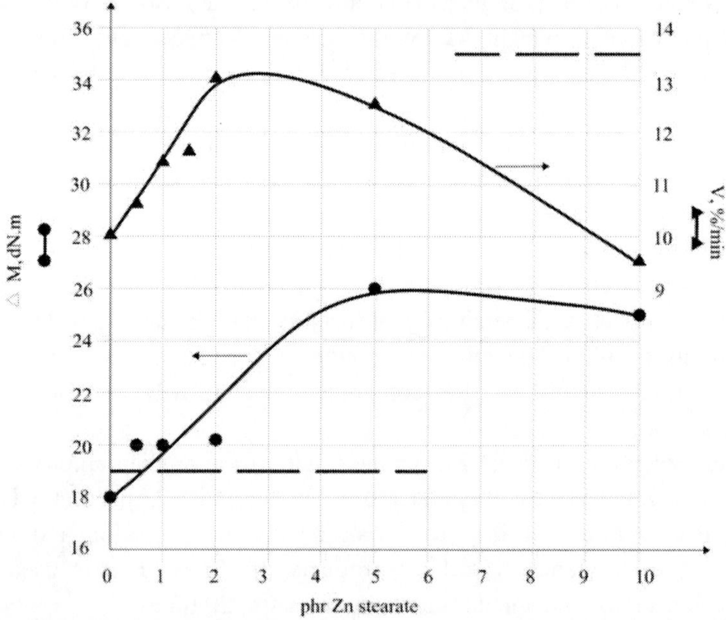

Figure 6. Dependence of ΔM = (MH – ML) /•-•/ and the curing rate V /▲-▲/ from the Zn stearate content.

The interaction between zinc oxide and stearic acid in a medium suitable to simulate a vulcanized system has been investigated [65] experimentally using vibrational spectroscopic technique. Confocal Raman micro spectroscopy revealed that at ambient temperature both components are phase-separated in the form of microcrystals. When the reaction temperature (80°C and above) is reached only zinc oxide is present in the form of particles while the stearic acid melts and gets molecularly dispersed within the rubber matrix. The analysis points to a core-shell structure of the reacting system: stearic acid diffuses to the surface of zinc oxide domains causing the shrinkage of the zinc oxide core and the formation of a shell of increasing thickness made of zinc stearate.

The study [66] includes a complete analysis of how sulfur crosslink formation is influenced by the presence of zinc monomethacrylate (ZMMA). Swelling experiments and chemical probe desulfuration studies confirm the changes in the distribution of crosslinks in ZMMA- versus ZnO-activated cure. The performance of ZMMA as an activator is also studied as a function of the loading and relative ratio of sulfur and accelerator. Finally, the increased efficiency of ZMMA activation is applied to various model formulations in order to demonstrate the potential for reductions in zinc content in a variety of applications.

ZINC STEARATE COULD SOLVE ONE MORE ENVIRONMENTAL PROBLEM

For both environmental and economic reason, there is broad interest in recycling rubber and in the continued development of recycling technologies [68]. The use of postindustrial materials is a fairly well-established and documented business. Much effort over the past decade has been put into dealing with of end-of-life tires from landfills and vacant fields. It is only in the last few years that more business opportunities for recycled rubber have come to the forefront. Reclaiming rubber has gained increasing interest, more so in Europe than in North America. In those areas much work has been done to refine the processes used. The major form of recycled rubber is still ground rubber. It is produced either by cryogenic, ambient, or wet grinding. The material is then used neat with sulfur/curatives, binders, or cements. The binders are normally moisture curable urethanes, liquid polybutadienes, or latex to produce items such as mats, floor tiles, and carpet under cushion.

Recycled rubber is still used as tire derived fuel, but less so than 10 years ago. Another outlet is as an additive to asphalt. Recycled rubber can be used in the plastics industry, for which much development is being done. Large particle size ground rubber or chips are used in civil engineering applications, landscaping, or artificial turf.

Part of waste tires is recycled and one of the obtained products is rubber flour. Its use as a filler for rubber compounds is acceptable in amounts up to 7 phr because a greater amount leads to deteriorating quality of the finished products.

In order to make possible the increasing of the amounts of rubber flour in the rubber compounds a range of methods for processing the surface of their particles were developed. These methods are not going to be examined herein because there is no use of salts of the stearic acid. What will be examined in more details is the modification of the surface of rubber flour by means of zinc stearate [69].

Detailed research regarding characterization of rubber flour of different trade names suggests that extraction with organic solvents (such as phenol, chloroform, and ethanol) of the rubber flour results in obtaining substances of various compositions among which non vulcanized rubber. This result confirms that a co-vulcanization between the non-vulcanized rubber and the fresh rubber in the basic compound could be expected, if only the appropriate conditions could be determined. First of all, it was stated that the sequence of adding rubber flour to the rubber itself is of great importance for the vulcanization features of the rubber compound and the properties of the resulting vulcanizates. It was proved that better results are obtained [69] if the compound of all the ingredients, except the rubber flour, is firstly prepared. Rubber powder is added at the end and the sulfur and the accelerator are also added after cooling of compound. It was analyzed the opportunity to improve the chemical interaction between the surface of the rubber flour and the rubber in the basic compound. Creation of such strong bonds is possible when the surface of the rubber flour is processed beforehand in a suitable manner in order to obtain those bonds. For the practical realization of this idea there are no convenient methods related to wetting the rubber flour, swelling and eventually carrying out processes of drying in case of solvent or water use. It has been suggested [69] that it would be appropriate to carry out such a processing of the surface of the rubber flour “in the dry” which means by powdering with suitable substances. The “in the dry” processing of the rubber flour could probably be realized, in practice, quite easily by mixing the rubber flour with the substances already defined in an appropriate vessel; by means of treatment in a fluidized bed; by means of stirring in an appropriate rotatable drum etc. For this purpose the influence on the vulcanization characteristics of a preliminary treatment in the dry of the rubber flour with different ingredients of the compound (zinc oxide and stearic acid, zinc stearate, sulfur and CBS accelerator) was examined. The basic compound contains SBR – 100, carbon black N-550 – 50, sulfur - 2, CBS – 1.2, rubber flour, powdered with zinc stearate - 0, 5, 10, 20 and 50. The results show that the vulcanization characteristics of the compounds at 160°C do not change considerably. The mechanical properties of the vulcanizates containing different types of rubber flour modified in such a manner are comparable (table 2).

The common conclusion is that the tensile strength smoothly decreases; the decrease of the elongation at break is lower; the Shore A hardness increases a bit; the residual elongation of the vulcanizates varies to a great extent.

And what would be the result if there are 20 phr rubber flour powdered in advance with zinc stearate and imported into the compound, which contains classic activators, zinc oxide and stearic acid? The basic idea of such a study is to attain a better acceleration of the vulcanization at the border level between the surface of the rubber flour and the rubber macromolecules in the basic compound. As a result of a number of investigations it has been stated that the zinc stearate does accelerate the vulcanization. It has been suggested that in such a system the vulcanization process will firstly start and will run out as quickly as possible at the border surface between the rubber flour and the macromolecules, and after that will take place within the main mass of the compound. That is how a smooth transition between the vulcanizate in the rubber flour and the vulcanizate in the main compound is achieved in relation to: the vulcanization rate; the type and density of the cross links; the mechanical characteristics of the system. Taking into account the results, it is to be concluded that no substantial difference is observed in the behavior of the vulcanizates vulcanized with zinc oxide as an activator in the basic compound and with zinc oxide and stearic acid respectively.

There is some very interesting information about the wasted tires worldwide in a publication of Tire Technology International edition from representatives of ETRA – a European organization dealing with tires and recycling of vulcanizates. From the information contained in this work, 1 billion tires go to waste on a world scale. EU data show that wasted tires in EU member states for 1996 were about 2.5 million tons, for 1998 almost the same, and for 2008 the quantity is expected to be about 3 million tons. It has been stressed that 100% recycling of wasted tires is possible and this would save up energy and materials.

STEARIC ACID AND ZINC STEARATE – RELEASE MOLD AGENTS

Good appearance is of great importance for rubber articles. It is the working surface of the mold which generally determines it. In order to achieve a smooth and brilliant rubber article, this surface is to be polished and chromium treated.

The mold complexity is also important as it determines the quick and safe removal of the article from the mold. The abovementioned is only one "side of the medal". After the long-lasting and continuous production processes onto the working surfaces, a fouling is obtained within each mold. For this reason, it is necessary to periodically clean the mold which upset the processing and is related with additional production expenses. In order to decrease the fouling of the molds, a variety of separating substances are used for spraying onto their working surfaces.

There are not many investigations published concerning these problems. Of course, it does not mean that there are no serious investigations carried out. Probably, all the resulting issues are qualified as business secret. The fact that some authors [71] named their publication as "Minimal Mold Fouling – a Miracle or Know-How?" is obviously not a coincidence. In 2014, the Second Conference Release and Clean took place in Germany [72] but presentations and abstracts still have not been published.

Table 2. Stress strain characteristics of vulcanizates containing rubber flour [69]

Property phr	0	5	10	20	50
Stress at 100% elongation, MPa	2.4	2.3	2.6	2.5	2.6
Stress at 300% elongation, MPa	13	12	13	11	10
Tensile strength, MPa	21	18	14	15	13
Elongation at break, %	415	500	330	380	380
Residual elongation, %	9	11	11	19	27
Hardness Shore A	65	66	70	70	71

In the process of production of rubber articles a lot of trade products designated to spray the working surfaces of the molds is used – ranging from soap solutions to silicone oils. However, these separating substances are not suitable for the production of elastomeric pharmaceutical closures and toys from a soft thermoplastic elastomer [73].

There are a lot of important requirements regarding the features of the elastomeric pharmaceutical closures [73 - 78]. Their widest application for them was founded to be butyl rubber (IIR) and its halogen based derivatives such as chlorine-butyl (CIIR) and brome-butyl rubbers (BIIR). The properties determining such an application are: low gas permeability, nontoxicity, fragmentation resistance, stable properties in the presence of acceptable antioxidants, and resistance towards medical products. These requirements, except the gas permeability, are met by natural rubber (NR) and by ethylene propylene triple co-polymer (EPDM). Vulcanizates from the NR have about 20 times higher gas permeability, and the vulcanizates from EPDM – 10 times higher, compared to the vulcanizates from IIR, CIIR и BIIR [77]. In order to obtain permission from the appropriate authorities for the production of elastomeric pharmaceutical stoppers more than 20 different physical, chemical and biological analyses were carried out in specialized laboratories. There is a strongly specified norm for each of the indices. The requirements concerning the content of such reducing substances, mercaptans, zinc ions, sulfates, chlorides, heavy metals are fairly high. In-depth investigations [74] of the influence of the composition and the vulcanization conditions on the content of these substances within an extract of the vulcanizate, based on butyl rubber and its halogen derivatives, were carried out.

Table 3. Consistency coefficient K (x10^3 Pa.s) of butyl rubber compounds with different additives

phr	0	3	5	6	8	10
Rubber compound	106	-	-	-	-	-
Glycerin	106	106	106	106	-	-
CR	106	98	98	98	98	-
Low molecular PE	106	102	79	-	-	79
Zn stearate	106	101	66	62	-	-
Dispergum C	106	95	72	70	-	-
Dispergum L	106	86	64	57	-	-
Butyl rubber without additives	90	-	-	-	-	-

It is known that concerning the rubber articles for medical purposes, pharmacy and food products, using carbon black is not allowed. It was stated [79] that liberation from 0.06 – 0.21 μg/l benzpyrene towards modeled media and food products does occur. By increasing the amounts of carbon black contained in the vulcanized rubber the migration of the benzpyrene, which is cancer genic, also increases. That is why mineral fillers are used for production of the elastomeric pharmaceutical stoppers – special marks kaolin, silica etc. As already stated stearic acid does not influence the dispersion of mineral fillers within the rubber and consequently is not included in such rubber compounds. However, in the case of the zinc salt of the stearic acid there is a great positive multifunctional effect. Zinc stearate significantly decreases the effective viscosity of rubber compounds from butyl rubber and from its halogen derivatives filled with kaolin and silica (table 3) [78]. The results in table 3 show that 5 phr zinc stearate decrease significantly the consistency coefficient K of the compound. This is very favorable for the production technology based on injection molding.

By the 1990s, we had developed a technology for the production of elastomeric pharmaceutical stoppers from chlorine-butyl rubber by injection molding. The content of zinc stearate in the rubber compound gave the stoppers an appearance of high quality; after the vulcanization the stoppers were removed very quickly and easily from the multiple mold and the fouling of the mold was reduced to a minimum. In the production of the stoppers from a compound without zinc stearate there was a need to remove the molds from the machine and to clean each of them for two weeks. When the production of the stoppers was from compounds with zinc stearate, the removal of the molds from the machine for cleaning purposes once lasted three months. Using such a compound, over 100 million stoppers designated for sealing vials for antibiotics were produced in Bulgaria. These results are from a personal production experience and have not been published.

CONCLUSION

In conclusion, it should be summarized that stearic is an important ingredient for rubber compounds and blends. For almost 100 years it has continued to be the unique dispergator for carbon black filled rubber compounds. On the basis of measurements of the rheological properties of the rubber compounds and depending on the stearic acid in them, specifying the optimal amount of the dispergator for the carbon black contained in the rubber has been developed and presented as part of this technical paper. Regarding the reinforcement of rubbers, research should be directed towards a fundamental understanding of the significance of stearic acid acting as a dispergator for the carbon black. It is expected that a universal dispergator for mineral fillers for use in the rubber industry would also be discovered. The stearic acid (used in common with the zinc oxide) is an important activator of sulfur accelerated vulcanization of non-saturated rubbers. The mechanism of sulfur vulcanization has not yet been explained in depth, but fundamental investigations were carried out on separated steps of this very important process. In that respect we estimate highly the publication of Italian scientists [65] in which are presented the results of investigations concerning the interaction in a modeled rubber medium between stearic acid and ZnO followed by obtaining zinc stearate. The heterogeneous nature of the reaction system has been confirmed by the use of contemporary methods of investigations. This is a great success and an important contribution towards the explanation of the vulcanization mechanism. Thus the results obtained from studies presented above on the role of zinc stearate as a vulcanization activator have been confirmed and explained. The use of zinc stearate as an activator, instead of stearic acid and zinc oxide, would reduce the amount of zinc within the rubber vulcanizates and consequently to decrease the environment pollution. The method presented herein of rubber flour "in dry" processing by zinc stearate and the addition to the rubber compounds of rubber flour in higher amounts (such as 20 phr) could have a positive ecological effect as well.

The present review does not claim exhaustiveness. We just tried to formulate essential tendencies concerning the role and the of stearic acid and zinc stearate for the purposes of rubber chemistry and technology. We noted the main tendencies of the forthcoming development of scientific and technological research in the field. Special attention was paid to the chances of success for stearic acid and zinc stearate to improve the environment and create better conditions for living on our planet.

REFERENCES

[1] Alliger, G; Sjothun, IJ. Vulcanization of Elastomers, 1964, p.131 Rheinhold Publishing Corporation, New York.

[2] Nikolinski, P. *Rubber Technology*, 1962, p.127 Edition Techniques, Sofia (in Bulgarian)

[3] Dogadkin, BA. Chemistry of Elastomers, 1972, p.243 Edition Chemistry, Moscow (in Russian).
[4] Kleemann, G. Compounds for Rubber Processing, 1984, p.143 Edition VEB Deutscher Verlag für Grundstoffindustrie, Leipzig (in German)
[5] Djagarova, E. *Rubber Technology*, 2000, 87, 74 Edition Artgrafik, Sofia (in Bulgarian)
[6] Donnet, JB; Custodero, E; Wang, TK. Atomic Force Microscopy of Carbon Black Aggregates. *Kaut. Gummi Kunstst.*, 1996, Vol.49, № 4, 274 – 281
[7] Guerle, L; Freakley, PK. A Comparison of Some Indirect Methods of Measuring Carbon Black Dispersion in Rubber Compounds. *Kaut. Gummi Kunstst.*, 1995, Vol.48, № 4, 260 – 269.
[8] Kim, PS; White, JL. Comparison of Black Incorporation and Development of Dispersion in Intermishing and Separated Counter-Rotating Rotor Internal Mixtures. *Kaut. Gummi Kunstst.*, 1996, Vol. 49, № 1, 10 – 17.
[9] Dimitrov, R; Sotirova, M; Gegova, E. Electrical Conductivity – Criterion for Determination of the Optimum Time for Homogenization of Black-filled Rubber Compounds. *Chem. Industry*, 1971, Vol. 43, №2, 59 – 61 (in Bulgarian)
[10] Le, HH; Ilisch, S; Hamann, E; Keller, M; Radusch, H-J. Effect of Curing Additives on the Dispersion of Kinetics of Carbon Black in Rubber Compounds. *Rubber Chem. Technol.*, 2011, Vol.84, № 3, 415 – 424.
[11] Persson, S. Measurement of the Carbon Black Dispersion by means of Light Microscopy. *Polymer Testing*, 1984, Vol. 4, № 1, 45 – 59.
[12] Lloyd, DG. Comparison of Lubricants and their Influence on the Rheological Properties of Rubber Compounds. *Materials Design*, 1991, Vol. 12, № 3, 139 – 146.
[13] Pomchaitaward, C; Manas-Zloczower; Feke, DL. Investigation of the Carbon Black Dispersion in Rubber. *Chem. Eng. Sci.*, 2003, Vol. 92, № 1, 17 – 24.
[14] Qi, Li; Feke, DL. Manas-Zloczower Effect of the shear stress on the Carbon Black Dispersion in Liquid Medium. *Powder Technology*, 1997, Vol. 92, № 1,17 – 24.
[15] Natov, M; Djagarova, E. Influence of Low Molecular Substances on the Viscosity of the Polymer Melts. *High Molec. Comp.*, 1966, Vol. 8, № 10, 1841 – 1847 (in Russian).

[16] Natov, M; Djagarova, E. Influence of Low Molecular Additives on the Viscosity of Polyethylene Melt. *Makromolek. Chemie*, 1967, Vol.100, 126 – 138 (in German)

[17] Natov, M; Djagarova, E. Changes in the Flow Properties of the Polypropylene Melt under the Influence of Small Amounts of Additives of Low Molecular Organic Substances. *Angew. Makromolek. Chemie*, 1968, Vol.2, 165 – 179 (in German)

[18] Natov, M; Djagarova, E. Influence of Small Amounts of Powdered Substances on the Flow Properties of the Polymer Melts. *Angew. Makromolek. Chemie*, 1968, Vol.2, 180 – 189 (in German)

[19] Natov, M; Abasova, J. On the Rheological Behavior of NBR and his Compounds. *Plaste Kaut.*, 1975, Vol.22, № 12, 962 – 969 (in German)

[20] Natov, M; Abasova, J. Rheological Properties of the 1,4 -cis-Polyisoprene and his Compounds. *Plaste Kaut.*, 1977, Vol.24, № 10, 699 – 706 (in German)

[21] Natov, M; Abasova, J. On the Rheological Properties of the Natural Rubber and his Compounds. *Plaste Kaut.*, 1978, Vol. 25, № 1, 27 – 34 (in German)

[22] Burmistr, MV; Ovtsharov, VI; Suhi, KM; Sokolova, LO; Gomza, JP; Djagarova, E. *Influence Modified Bentonit on the Properties of the Polyisoprene Rubber Compounds Polym. J.*, 2007, Vol. 29, № 1, 16 – 22 (in Ukrainian)

[23] Zheleva, D. PhD Thesis: "Rheological Properties of Filled Rubber Compounds with Technological Additives", 2003, University of Chemical Technology and Metallurgy, Sofia (in Bulgarian)

[24] Djagarova, E; Jeleva, D; Zdravkov, Z. One Possibility to Enlargement of the Information by Measuring with Plasticorder Brabender. *J. Chem. Technol. Metallurgy* (UCTM), 2002, vol.37, №5, 71 – 78 (in French)

[25] Jeleva, D. Influence of Some Additives on the Rheological Properties of Filled Rubber Compounds. Proceedings of Conference on Processing and Application of Polymers TECHNOMER, 2001, 139-143, Chemnitz (Germany) (in English)

[26] Zheleva, D. Comparison of the Dispersing Capability of the Stearic Acid and Zinc Stearate Based on the Rheological Properties of Rubber Compounds. Proceedings of Conference on Processing and Application of Polymers TECHNOMER, 2011, 137-142, Chemnitz (Germany) (in English)

[27] Zheleva, D. Method for Determination of the Optimum Amount of the Dispergator in Rubber Compounds. Proceedings of National Conference

"Science and Production Problems in Rubber Industry", 2010, Zarevo (Bulgaria) (in Bulgarian)

[28] Zheleva, D. Investigation of Type Carbon Black on the Stearic Acid Dispersion Effect in Filled Rubber Blends. Proceedings of Conference on Processing and Application of Polymers TECHNOMER, 2003, 112-117, Chemnitz (Germany) (in English)

[29] Mark, J; Erman, B; Eirich, F. Science and Technology of Rubber, 2005, Elsevier Press Inc., 3th Edition, 205, 237.

[30] Chhabra, R; Richardson J. Non-Newtonian Flow in the Process Industries, 1999, Edition Butterworth Heinemann, Oxford Auckland Boston Johannesburg Melbourne New Delhi , 43-46.

[31] Popovic, RS; Plavsic, M; Popovic, RG; Ilic N. Correlation between Viscosity Measured by Capillary Rheometer and Mooney Viscometer on Testing Rubber and Rubber Mixes. *Kaut. Gummi Kunstst.*, 1991, vol.44, №7, 702 – 709.

[32] Zuo, M; Zhing Q. Correlation between Rheological Behavior of Filled Rubber Blends and their Properties and Structure. *Science China, Series B, Chemistry*, 2008, vol.51, №1, 1 – 7.

[33] Kuleznev V. *Polymer Blends*, 1966, Edition Chemistry, Moscow (in Russian)

[34] Zheleva D. An Attempt for Correlation between Mooney Viscosity and Rheological Properties of Filled Rubber Compounds. *J. Chem. Technol. Metallurgy (UCTM)*, 2013, vol.48, №3, 241 – 246.

[35] Ahn, SH; Kim, SH; Lee, SG. Influence of the Stearic Acid on the Silica Distribution in Rubber Compounds. *J. Appl. Polymer Sci.*, 2004, vol.94, №2, 802 – 811.

[36] Kosmalska, A; Zaborski, M; Slusarski, L. Effect of ZnO on the Silica Dispersion in Rubber Mixtures. *Macromol. Symp.*, 2003, vol.194, 269 – 277.

[37] Mukhopadhyay, R; De, SK. Effect of Vulcanization Temperature and Different Fillers on the Properties of Efficiently Vulcanized Natural Rubber. *Rubber Chem. Technol.,* 1979, vol.52, №2, 263 – 277.

[38] Choi, SS; Park, B-H; Song H. Influence of Some Accelerators on the Silica Distribution in Rubbers. *Polymer Adv. Technol.*, 2004, №15, 122 – 133.

[39] Tricas, N; Agullo, N; Borros S. Surface Modification of Carbon Black by Plasma Techniques on the Vulcanization Reaction. *Kaut. Gummi Kunstst.*, 2009, vol. 62, №3, 282-289.

[40] Jönnson, U; Malmquist, M; Rönnberg, N. Modification of the surface stress of the Silica. *J.Biochem.*, 1985, vol. 227, №2, 363 – 372.

[41] Osswald, K. Influence of the Additives on the Silica Distribution in NBR/NR Blends. Proceedings of Conference on Processing and Application of Polymers TECHNOMER, 2011, V 3.2, Chemnitz (Germany) (in German)

[42] Calzonetti, JA; Christopher, JL. Patents of Charles Goodyear: his International Contribution to the Rubber Industry. *Rubber Chem. Technol.,* 2010, vol.83, №3, 303 – 321.

[43] Serier, J-B. Histoire du caoutschouc, 1993, Edition Desjonquieres, Paris, 83 - 86 (in French)

[44] Meyer, KH. High Molecular Compounds 1940, Edition Akademische Verlaggeselschaft, Leipzig-Berlin, 10 -14 (in German)

[45] Heideman, G. PhD Thesis: "Reduced Zinc Oxide Levels in Sulfur Vulcanization of Rubber Compounds", 2004, University of Twente, The Netherlands.

[46] Kresja, MR; Koenig, JI. "The Nature of Sulfur Vulcanization" in Elastomer Technology Handbook, 1993, CRC Press, New Jersey, 5 – 8.

[47] Kohjiya, S; Tosaka, M; Furutani, M; Ikeda, Y; Toki, S; Hsiao, BS. Role of Stearic Acid in the Strain-Induced Crystallization of Cross Linked Natural Rubber and Synthetic cis-1,4-Polyisoprene. *Polymer*, 2007, vol.48, 3801 – 3808.

[48] Chukwn, MN; Madufor, IC; Ayo, MD; Ekebafe, LO. Effect of Stearic acid Level on the Physical Properties of Natural Rubber Vulcanizate. *Pacific J. Sci. Technol.*, 2011, vol.12, №1, 344 – 350.

[49] Garreta E.; Agullo N.; Borros S. The Role of the Activator during the Vulcanization of Natural Rubber using Sulphenamide Accelerator Type. *Kaut. Gummi Kunstst.*, 2002, vol. 55, №3, 82-91.

[50] Coran, AY. Vulcanization. Part III. Rapid Methods for Characterizing Rubber Networks. *Rubber Chem. Technol.,* 1964, vol.37, №3, 668 – 682.

[51] Ding, R; Leonov, AI; Coran, AI. A Study of the Vulcanization Kinetics of an Accelerator-Sulfur-SBR Compound. *Rubber Chem. Technol.,* 1996, vol.69, №1, 69 -81.

[52] Heideman, G; Noordermeer J. Data R. Reduced ZnO Levels in Sulfur Vulcnization of Rubber Compounds. *Tire Techn. Intern., Annual Review*, 2004, 22 – 31

[53] Chapman, AV. "Safe Rubber Chemicals: Reduction of Zinc Levels in Rubber Compounds" IRC, 2005, Maastricht.

[54] International Zinc Association (IZA): Zinc in the Environment, 1997, Brussels.

[55] World Health Organization (WHO): Environmental Health Criteria 221, Zinc, 2001, Genève.

[56] Schröter, A. Zinc Oxid in Rubber Industry. *Tire Technol. Intern.*, 2005, №1, 68-69.

[57] Brodska, A; Hrdlicka, Z; Kuta A. Effect of ZnO with Different Specific Surface Area on the Cure Characteristic and Mechanical Properties of Carbon Black Filled NR/SBR Compound. *Kaut. Gummi Kunstst.*, 2012, vol. 65, №9, 45 – 52.

[58] Guzman, M; Vega, B; Agullo, N; Giese, U; Borros S. Zink Oxide versus Magnesium Oxide Revisited. Part I. *Rubber Chem. Technol.*, 2012, vol.85, №1, 38 – 55.

[59] Guzman, M; Vega, B; Agullo, N; Borros S. Zink Oxide versus Magnesium Oxide Revisited. Part II. *Rubber Chem. Technol.*, 2012, vol.85, №1, 56 – 67.

[60] Tipova, N. PhD Thesis "Investigation of Zinc Stearate as Activator of the Accelerated Sulfur Vulcanization with a view to Zinc Reducing in the Rubber Compounds", 2006, University of Chemical Technology and Metallurgy, Sofia (in Bulgarian).

[61] Tipova, N. Influence of the Zinc Stearate as an Activator for the Accelerated Sulphur Vulcanization aiming the Reduction of Zinc Level in the Vulcanizates. *Proceedings of Conference on Processing and Application of Polymers TECHNOMER*, 2007, 112-127, Chemnitz (Germany) (in English).

[62] Djagarova, E; Tipova N. On the Possibility for Reduction of the Zn Amount in Rubber Compounds. *Proceedings of German Rubber Conference*, 2006, Nurnberg (Germany) (in German)

[63] Djagarova, E; Tipova, N; Iliev Pl. Influence of the Surface Modification of the Rubber Flour on the Properties of the Rubber Compounds. *Proceedings of Conference on Processing and Application of Polymers TECHNOMER*, 2005, 112-127, Chemnitz (Germany) (in German)

[64] Helaly, FM; Sabbagh, SH; Kinawy, OS; Sawy S.M. Effect of Synthesized Zinc Stearate on the Properties of Natural Rubber Vulcanizates in the Absence and Presence of some Fillers. *Materials Design*, 2011, vol.32, 2835 – 2843.

[65] Musto, P; Larobina, D; Cotugno, S; Straffi, P; DiFlorio, G. Confocal Raman Imaging, FTIR Spectroscopy and Kinetic Modeling the Zinc

Oxide/Stearic Acid Reaction in a Vulcanizing Rubber. *Polymer*, 2013, vol. 54, 685 – 693.

[66] Monsallier J.-M. Using Zinc Monomethacrylate: Active Accelerated Sulfur Vulcanization and Reduce Zinc Loading. *Kaut. Gummi Kunstst.*, 2009, vol. 62, №11, 45 – 52.

[67] Chapman, A; Johnson T. The Role of Zinc in the Vulcanization of Styrene-Butadiene Rubber. *Kaut. Gummi Kunstst.*, 2005, vol. 58, №7, 358 – 361.

[68] Myhre, M; Saivari, S; Dierkes, W; Noordermeer J. Rubber Recycling, Chemistry, Processing, and Applications. *Rubber Chem. Technol.*, 2012, vol.85, №3 , 408 – 449.

[69] Djagarova, E; Tipova, N; Iliev Pl. Influence of the Surface Modification of the Rubber Flour on the Properties of the Rubber Compounds. *Gummi Fasern Kunstst.*, 2006, vol. 59, №6, 380 – 385.

[70] Ziegler, E;Macher, J; Gruber, D; Pölt, P; Kern W. Investigation of the Influence of Stearic Acid on Rubber-Brass Adhesion. *Rubber Chem. Technol.*, 2012, vol.85, №2, 264 – 276.

[71] Utz, R; Henzel, M; Sprenger S. Minimal Mold Fouling – a Miracle or Now How? *Kaut. Gummi Kunstst.*, 1995, vol. 48, №2, 104 – 111.

[72] *Kaut. Gummi Kunstst.*, 2014, vol. 67, №1.

[73] Böhm P. Optimization of the Article by means of TopoSeal in Molds - a Rapid Road to Effective Removing. *Proceedings of Conference on Processing and Application of Polymers TECHNOMER*, 2011, 112-127, Chemnitz (Germany) (in German)

[74] Djagarova, E,; Gegova, E. Influence of the Formulation and Vulcanization Conditions on the Amount of the Mercaptans and Reducing Substances in an Extract of Butyl Rubber Vulcanizates. Polymers, *Proceedings of the Institute for Rubber Industry Sofia*, 1984, vol.2, №16, 69 – 76 (in Bulgarian)

[75] Jansen R. Butyl Rubber in Pharmaceutical Applications: Experience, Knowledge and Trends. *Kaut. Gummi Kunstst.*, 2013, vol. 66, №1, 49 – 54.

[76] Harmsworth, N. Special Properties of the Butyl Rubber. *Kaut. Gummi Kunstst.*, 1995, vol. 48, №1, 38 – 45.

[77] Bochossian, T; Djagarova, E; Dimitrov V. Application of the Gas Chromatography for Measuring of the Vulcanizates Gas Permeability. *Proceedings of the* 5th *Conference on the Problems in the Production of the Rubber Articles*, Sofia, 1987, 156 – 161 (in Bulgarian).

[78] Djagarova, E. Some Characteristic of Chlorobutyl Rubber Compounds Rheological Properties. *Proceedings of the 7th International Conference on Mechanics and Technology of Composite Materials*, Sofia, 1994, 51 – 56 (in English).

[79] Stankevitch, KI. Manual on the Hygiene of Polymer Application, 1984, Edition Health, Kiev, 107 – 109 (in Russian)

In: Stearic Acid
Editors: Yunfeng Lin and Qiang Peng
ISBN: 978-1-63463-172-3

Chapter 3

STEARIC ACID IN BIOMEDICAL SCIENCE

Ting Zhang[1], Qiang Peng[2,*] and Yunfeng Lin[2,†]

[1]Key Laboratory of Drug Targeting and Drug Delivery Systems, Ministry of Education, West China School of Pharmacy, Sichuan University, Chengdu, China

[2]State Key Laboratory of Oral Diseases, West China Hospital of Stomatology, Sichuan University, Chengdu, China

ABSTRACT

Stearic acid, an endogenous long-chain saturated fatty acid, is known as nontoxic, biocompatible, biodegradable and cost effective. Consequently, it has been widely used in biomedical science for years. Traditionally, stearic acid and its stearates, often magnesium stearate, are widely used as lubricants. Besides, stearic acid, an amphiphilic molecule with a hydrophobic group at the "tail" of the molecular structure and hydrophilic groups at the "head", is a surface active substance with extensive application. It could be used as emulsifier in emulsion, solubilizer for insoluble or slightly soluble substance and defoamer in some pharmaceutical industry. Stearic acid and its derivatives could also be used for cream and suppository preparation due to their appropriate physicochemical properties, such as melting point and acidity.

* Qiang Peng: State Key Laboratory of Oral Diseases, West China Hospital of Stomatology, Sichuan University, Chengdu, 610041, China. E-mail: lijm2002@163.com.

† Yunfeng Lin: State Key Laboratory of Oral Diseases, West China Hospital of Stomatology, Sichuan University, Chengdu, 610041, China. E-mail: yunfenglin@scu.edu.cn.

Moreover, stearic acid is a common constituent in enteric coating for its plasticity and stability in stomach. Occasionally, it is used as binder agent especially in melt pelletization. In addition to be a traditional biomedical material, which has been used for thousands of years, stearic acid also plays an important role in newly developed drug delivery system, like microparticle dispersion system which is usually used for sustained drug delivery. Stearic acid is a vital ingredient candidate for nanoparticle and microsphere preparation.

INTRODUCTION

As we know, biomedical science is a vital and necessary field in our daily life, which is aimed at solving medical problems by biotechnology.

In biomedical science, biomaterial is often used for drug carrier or pharmaceutical auxiliary, among which safe and nontoxic ones have attracted more attention of scientists.

Traditionally, natural oil has been used as soap and lubricants in our life for more than 1400 years. Since fatty acids were proved to be a component of oil in 1823, the application of fatty acids has been developed quickly.

In biomedical science, fatty acids have been investigated as carriers or for the development of drug delivery systems as they are considered to be endogenous, cost-effective and biocompatible. Stearic acid is a long-chain of 18 carbon atoms saturated fatty acid atoms which is the main component of fat. Initially, it was used for candle production.

Afterwards, stearic acid was replaced by paraffin wax in this field, which, however, did not reduce the consuming amount of stearic acid. As it is nontoxic, tasteless and has relatively high melting point and fine plasticity, its application is increasingly extensive, and biomedical science included.

Stearic acid is initially widely used as lubricant in medicine industry [1]. Subsequently, because of its surface activity resulted from amphiphilic molecular structure, stearic acid and its derivatives are also used as solubilizer [2], emulsifier [3], defoamer [4], and the basis of cream [5] or suppository [6]. Its other applications, such as enteric coating material [7], binder agent [8] and some other novel usages [9-11] are also described in this chapter.

A Common Usage in Pharmaceutical Manufacturing – Lubricant

Generally, tablet is a common pharmaceutical dosage form in clinical and prepared mainly in granulation tabletting and powder compression technique. Regardless of the tabletting methods, various auxiliary substrates are needed, among which lubricant is necessary. Lubricant mainly has the following three functions during the tabletting process:

- Contributing to adding ingredients and ejecting tablets smoothly;
- Preventing sticking of materials to the punch;
- Lowering the friction force between granules or granules and die hole, because of which smooth and uniform tablets become possible.

The mechanism of lubricant function is complex and still not very clear. Briefly speaking, lubricants are able to improve the surface characters of particles (small in size and large in surface area). In detail, there are three main aspects as follows:

1 Improving the electrostatic distribution of the particle surface;
2 Decreasing the surface roughness of the particles to reduce friction force;
3 Improving the selectivity of gas adsorption and reducing the Van Der Waals' force between particles.

Lubricants are usually classified into three types, hydrophobic and hydrophilic lubricants, and glidant, among which hydrophobic lubricant is most frequently used. It includes fatty acids, alcohols, esters and salts of fatty acids and oils. Magnesium stearate is the most widely used lubricant in pharmaceutical industry, such as tablet and capsule preparation. It is in the form of white exquisite powder with low bulk density and high specific area, so it is easy to be mixed with granules and adhere to granule surface, which is in favor of decreasing friction force. With fine lubricity, magnesium stearate could lower the ejection force significantly and avoid punch sticking, contributing to improving the appearance of tablets. The contact angle, a parameter of wetting ability, of magnesium stearate is as high as 121°. For the high hydrophobicity of magnesium stearate, improper use may lead to too high hydrophobicity of tablets, influencing its wetting by water, disintegration and subsequently drug dissolution. Moreover, it also has effect on the rigidity of tablets. Therefore, it is necessary to be used in proper concentration, which is usually in the range of 0.1% - 1.0%.

For example, the formulation of venoruton tablets contains venoruton 0.1g, microcrystalline cellulose 0.03g, starch 0.01g, Aluminum magnesium powder 0.07g and magnesium stearate 0.004g.

Stearic acid, also known as octadecanoic or stearophanic acid, is also widely used as lubricant, only second to magnesium stearate. The contact angle of stearic acid is 98°. It exhibits good lubrication ability and relatively high hydrophobicity, both of which, however, are weaker than that of magnesium stearate. The stearic acid in market is usually in big pieces and it is difficult to be grinded directly. So it is often broken into small ones firstly and then grinded with talcum powder. The common concentration of stearic acid as lubricant in formulation is 1-3%. For instance, in paracetamol tablet preparation, paracetamol 80.0 kg, stearic acid 12.4 kg and 10% ethycellulose solution in ethanol are mixed for granulation, followed by tabletting with 30 kg magnesium stearate. Per piece containing 0.2 - 1.0 g paracetamol.

More examples can be found in references. In two traditional sodium aspirin tablet formulations, stearic acid and magnesium stearate are used as lubricant in the concentration of 1.5% and ~0.3%, respectively [12].

According to Chinese Pharmacopoeia, the quality standards of tablets include appearance, the difference in content and weigh, rigidity and friability, disintegration, dissolution and releasing rate. Considering the function of lubricant, it is not hard to see that lubricants could affect almost all the above parameters in tablets manufacturing, and stearic acid and magnesium stearate is no exception. For example, RC Rowe, et al. studied the effect of some lubricants on the adhesion of hydroxypropyl methyl cellulose film coatings and found that, the addition of magnesium or calcium stearate to the tablet could decrease the adhesion of film coatings to tablet surfaces, but the addition of stearic acid caused a significant increase [13].

Researchers also studied the effect of lubricants on dissolution rate of the active ingredient, which indicated that the more commonly used hydrophobic lubricants (magnesium stearate, aluminum stearate, calcium stearate, glyceryl monostearate, stearic acid) could decrease the effective drug-solvent interfacial area and thereby decrease the rate of dissolution of the drug, while water-soluble lubricants (sodium oleate, sodium lauryl sulfate, polyethylene glycol 4000 and talc) did not effect the dissolution rate [14, 15].

Meanwhile, Shibata et al. [16] elaborated the reason for the effect of magnesium stearate concentration on ethenzamide dissolution rate by SEM.

With the mixing time of 0.5% magnesium stearate and glass beads increased from 1 min to 30 min, a film of magnesium stearate was formed on the surface of the glass beads. And other reports proved that [17, 18], this hydrophobic film of magnesium stearate on the surface can reduce surface wettability, which subsequently reduces not only water penetration into a granule or tablet but also contact between drug and solvent [14], resulting in a decrease of the surface area directly contacting the solvent and a decrease in drug dissolution rate.

In Late SG et al.'s study, the effects of disintegration-promoting agent, lubricants and moisture treatment on optimized fast disintegrating tablets were studied elaborately, in which magnesium stearate was chosen as a lubricant at 1.5% concentration as it gave optimum hardness value with low disintegration time compared with several other lubricants, i.e., talc, stearic acid, glycerol dibehenate, L-leuine [19].

Moreover, the types and concentrations of lubricants have some other effects on the tabletting progress and the tablet products [20, 21].

Apart from magnesium stearate, some other stearate can also be used as lubricant, such as zinc, calcium and sodium stearate.

In addition, as an acid auxiliary substrate, stearic acid is not suitable for alkaline drug preparations to avoid chemical reaction.

STEARIC ACID – AN ANIONIC SURFACTANT

Because of the long chain carboxylic structure, stearic acid can be classified as surfactant. The science of surfactant, i.e., surface-active agent, has a long tradition and is prevailing in our daily life, as figure 1 shows [22].

With hydrophobic group at the "tail" of the molecular structure and hydrophilic groups at the "head", surfactant is amphiphilic. Their hydrophilic groups are often polar ones, such as carboxylic acid, sulfonic acid, sulfuric acid, or amine group and their salts, etc. And their hydrophobic groups are always nonpolar hydrocarbon chain with more than 8 carbon atoms.

Surfactants can be classified by various ways, and the most common classification method is based on the feature of molecular composition and the dissociative nature of polar groups.

The four basic kinds of surfactants are described as anionic, cationic, nonionic and amphoteric surfactants.

Stearic acid, as well as its salt and esters, has a negatively charged hydrophilic group, $RCOO^-$, so it belongs to anionic surfactants.

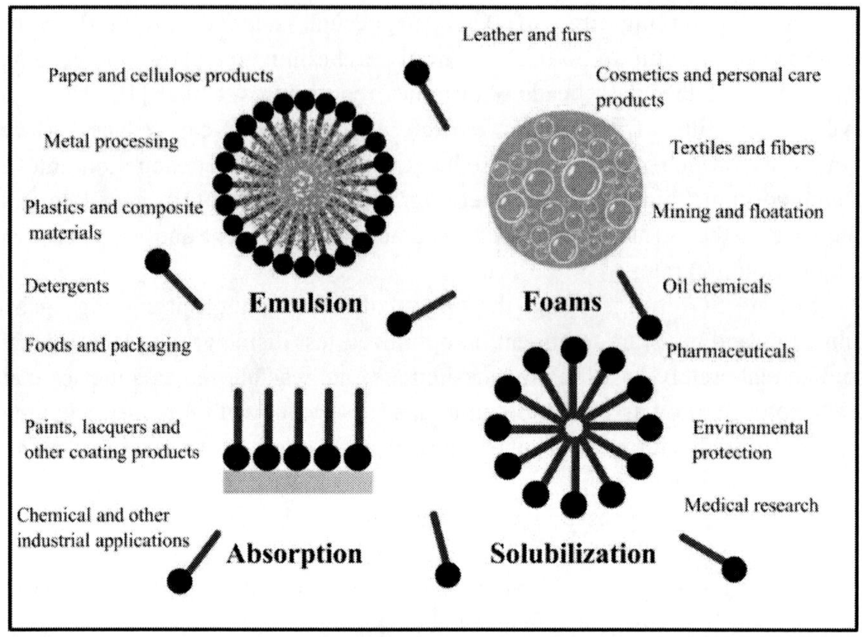

Figure 1. Some important, high-impact areas of surfactant applications.

Table 1. HLB ranges and their general areas of application

HLB Range	General Application
2-6	W/O emulsion
7-9	Wetting
8-18	O/W emulsion
3-15	Detergency
15-18	Solubilization

Hydrophile-lipophile balance (HLB) is an important parameter to evaluate the comprehensive affinity of hydrophilic and hydrophobic groups to oil or water. Experientially, the HLB value is defined in 0 - 40. The higher the HLB value, the stronger the hydrophility will be, and vice versa.

There is a close relationship between HLB value of surfactants and their application as Table 1 shows [23].

The HLB value of stearic acid is 17.0 and would vary with esterification or salification. Because of the surfactant features discussed above, stearic acid and its esters are widely used in pharmaceutics.

Emulsifier

Emulsification is a common method in pharmacy and surfactants can be used as emulsifier. Emulsion is a heterogeneous liquid dispersing system, in which at least one immiscible liquid dispersed in another in the form of liquid droplets, usually with a diameter of <0.1 mm. The liquid in small drops is called dispersed phase or internal phase, and the other one is called continuous phase or external phase. It mainly consists of three necessary components, water phase, oil phase and emulsifier. Based on the nature of surfactants and phase ratio, emulsion is usually described as oil-in-water (O/W) or water-in-oil (W/O), and multiple-emulsion is also possible, such as oil-in-water-in-oil (O/W/O) or water-in-oil-in-water. For emulsion is thermodynamically unstable system, four phenomena are commonly observed in this system, i.e., creaming, phase inversion, rancidification, coalescence and breaking. For enough shelf life of drug in emulsion, the stability of this preparation must be kept in mind when formulation designation. Emulsifier plays an essential role in emulsion stability. Firstly, it can reduce the surface tension effectively, contributing to the emulsion droplets formation, the newborn surface expanding and consequently maintaining the emulsion in a certain degree of dispersion and stability. Secondly, with the help of surfactants, stable emulsion could be prepared consuming less energy, even by simple shaking or stirring method, which is significant for industry manufacturing.

Stearic acid and stearates, as anionic surfactants, are widely used in emulsion preparation and study [3, 24, 25].

For example, the emulsion aerosol where stearic acid is used as emulsifier could be prepared by the method described below: All ingredients, stearic acid 9.5%, cocoa butter 2.0%, Cetyl alcohol 1.0%, glycerol 2.0%, potassium hydroxide 1.7%, sodium hydroxide 3.0%, water 83.5%, are heated and emulsified to obtain O/W emulsion; Drug could be added into the emulsion, oil phase, or water phase, as the case may be. The final aerosol product contains 5% propellants and 95% emulsion described above.

Bárány E et al. have studied the influence of stearic acid and several stearates, as emulsifiers in cream, on normal as well as on irritated skin [3]. Their results highlighted the possibility of absorption of these emulsifiers into the lipid bilayer, which increased trans-epidermal water loss (TEWL) in normal skin and decreased TEWL in damaged skin.

Although some stearates could be deleterious when used as emulsifiers in some instances according to previous study, the toxicity is largely limited to higher concentration [26].

Solubilizer

When the concentration of surfactant in water reaches certain degree, the solubility of some inherently insoluble or slightly soluble substance would increase greatly, resulted in transparent colloidal solution. This process is called solubilization and the surfactant is called solubilizer, while the aggregated units composed of a number of surfactant molecules are called micelles and the concentration at this critical point is named as critical micelle concentration (CMC). The interior of micelle is a tiny nonpolar hydrophobic space arranged by lipophilic groups, while the external is a polar region formed by the hydrophilic groups. Since the size of micelle is small enough in that of colloidal solution range, the solution appears transparent after drug is solubilized by micelles, resulting in the increased solubility. The shape of micelles are various, but mainly in five proposed ones as figure 2 shows [27], namely spherical, lamellar, inverted or reversed, disk, and cylindrical or rodlike. Based on various molecular structures, the locations for solubilization of additives in micelles are different as figure 3 shows [28].

Nonpolar additives are intimately associated with the core of the micelle (Figure 3a), while slightly polar materials are usually located in what is termed the "palisades layer" lying between the hydrophobic core and hydrophilic outer layer (Figure 3b). As shown in figure 3b, the orientation of these molecules is radial, with the hydrophobic tail closely associated with the micellar core and the hydrophilic head located intimately in the outer layer. What is more, besides additives located in the micellar core and the core-palisades boundary region, they may also distribute completely in the palisades region (Figure 3c) and on the outer surface of micelles (Figure 3d).

In micelles composed of some nonionic surfactants like polyoxyethylene derivatives with polar head group, the additives are preferentially located deep within the palisades layer (Figure 3c), whereas in case of ionics, the materials solubilized in this system may be intended to situated on the micellar surface (Figure 3d).

Insolubility of drugs is a great obstacle for drug function in clinical remedy and solubilization has been an important technology for solving this problem in biomedical science. A solubilized oral formulation could enhance the oral bioavailability of insoluble drugs and is an optimal choice for patients who can not take capsules or tablets.

While a solubilized injection formulation makes it possible for poorly water-soluble molecules used by injection, contributing to functioning more quickly than solid dosage forms administrated by non parenteral pathways.

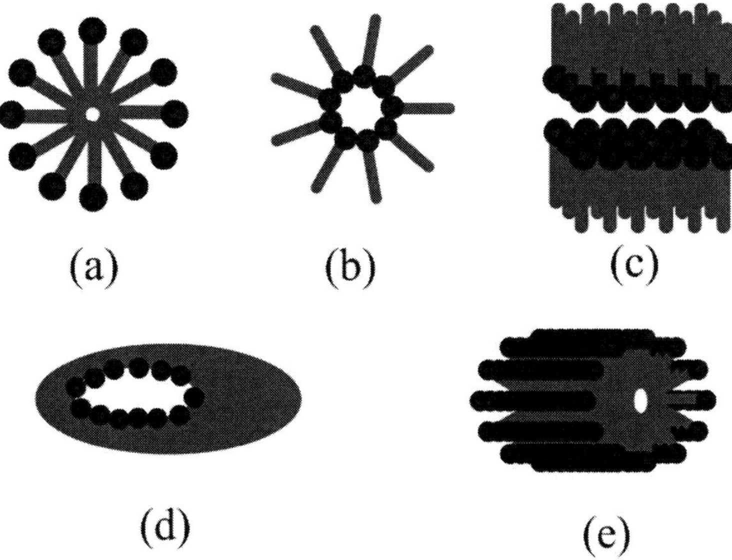

Figure 2. Five proposed shapes of micelle:(a) spherical; (b) inverted (or reversed); (c) lamellar; (d) disk; (e) cylindrical or rodlike.

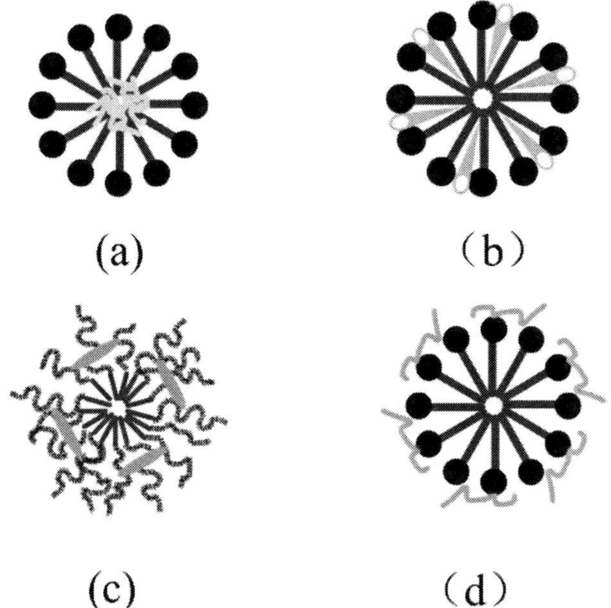

Figure 3. Locations for the solubilization of additives in micelles: (a) micelle core; (b) core-palisades interface; (c) surface region (nonionics); (d) on micelle surface (ionics).

As a solubilizer, stearic acid or its derivative can be found widely in biomedical science and technology. Lavasanifar A et al. [2] studied several solubilizing excipients, several mono-stearate included, for solubilizing water-insoluble drugs in oral and injectable formulations, which indicated that mono-stearic acid ester of PEG 400 and 1750 could be used as solubilizer for oral preparations. In another study, poly(ethylene oxide)-block-poly(N-hexyl stearate-Laspartamide) (PEO-b-PHSA) was synthesized to form block copolymer micelles by solvent evaporation method which was used to load amphotericin B (AmB) [29]. The degree of stearic acid substitution in PEO-b-PHSA molecules could be changed to increase the encapsulation efficiency of AmB and decrease the membrane disturbing effect named hemolysis, while its *in vitro* antifungal activity could be maintained at the same time.

Their results indicated that PEO-b-PHSA micelles with a high level of stearic acid side chain substitution, at 50% and 70%, can solubilize AmB effectively, reduce its hemolytic activity yet retain its potent antifungal effects.

In addition to traditional pharmaceutical preparations, stearic is also important in novel biomedical technology, gene medicine. Stearic acid grafted chitosan oligosaccharide (CSSA) is recently used as a promising gene carrier, [30-33] which is synthesized via the reaction between the carboxyl group of stearic acid with the amino groups of chitosan in the presence of EDC [34].

This novel molecule could self-aggregate to form micelle with low CMC in aqueous solution and condense DNA into nanoparticles easily, which exhibits good nuclease degradation protection ability and excellent internalization into cells. Jian You et al. prepared paclitaxel loaded CSSA micelles with about 40 nm diameter. With high degree of substitution, the micelles presented specific spatial structure with multiple hydrophobic "minor cores", contributing to micelles internalization into cancer cells and accumulation in cytoplasm. And the pH sensitivity of micelles is in favor of drug delivery by endocytosis and thus the cytotoxicity of paclitaxel encapsulated in micelles was enhanced greatly, which confirmed that the CCSA micelles are a promising vehicle for targeting therapy of anticancer drugs with a cytoplasmic molecule target [35].

As to the novel carrier material, Peng Hu et al. have evaluated the genotoxicity of stearic acid grafted chitosan oligosaccharide nanomicelles by *in vitro* and *in vivo* studies and found that, significant genotoxicity of high doses of stearic acid grafted chitosan oligosaccharide nanomicelles exhibited both *in vivo* and *in vitro*, the possible mechanism of which could partially be oxidative stress. However, studies on the field are limited and more researches should be done to evaluate this new biomaterial properly [36].

Defoamer

The agent used to prevent or mitigate foams is called defoamer. Foaming can be observed in solutions containing surfactants and/or under vigorous stirring, which is proved to be a great trouble. Defoamer is a kind of surfactant with weak surface tension and low hydrophility, whose HLB value is usually in the range of 1~3. As early as mid-20th century, stearic acid was frequently used to improve antifoam formulations [4, 37, 38].

Besides, stearate is also a potent antifoam agent. For instance, Ming Li Shi et al. studied the relationships between antifoaming potential and the number of hydrophilic groups as well as HLB of polyoxyalkylene ethers and its stearates. Their results confirmed that the distearate of polyoxyalkylene presented higher antifoam ability [39].

Stearic Acid – A Substrate in Semisolid Medicament

Semisolid medicament is characterized by soft, which can flow or deform by mild force or under body temperature. It is a user-friendly form, extruded easily and coated uniformly, which is widely used for external applications, such as skin, ocular region, nasal cavity, vagina, anus administration, etc. Cream and suppository are two common semisolid dosage forms and stearic acid is an important ingredient of their bases.

Cream

The semisolid emulsions which may be oil-in-water (O/W) or water-in-oil (W/O) type is defined as cream, in which drugs dissolved or dispersed in emulsion bases. Cream is composed of three parts, i.e., water phase, oil phase and surfactants. Because of its hydrophobicity, among others, stearic acid is frequently used as oil phase component.

The specific bases formulation was determined based on the nature and purpose of certain preparations. Table 2 provides three cream bases containing stearic acid and stearates [40].

Take NO.2 for example. In this formulation, stearic acid, liquid paraffin and albolene serve as oily phase as well as viscosity modifier.

Table 2. Four cream bases containing Stearic Acid

Prescription Composition	NO.1	NO.2	NO.3
Stearic Acid (g)	170	130	180
Liquid Paraffin (ml)	250	54	
Lanum (g)	20		
Triethanolamine (ml)	20	6.6	20
Glycerol (g)	50	100	50
Methylparaben (g)	1		
Ethylparaben (g)	1	1	
Glycerin monostearate (g)		38	50
Potassium Hydroxide (g)	7		
Albolene (g)		8	20
Paraffin (g)			50
Benzoic Acid (g)			1
Tween - 80 (g)			
Span - 80 (g)			
Adding Water to (ml)	1000	1000	1000

Table 3. Final composition of herbal cream (formulation F6) ([5])

Ingredients	Formulation %w/w in grams
Aq. Extract of Panax ginseng	5.0
Aq. Extract of Calendula officinalis	5.0
Aq. Extract of Arnica Montana	2.0
Aq. Extract of Clerodendrum	1.0
indicum	
Aloe vera	3.0
Rose hip oil	4.0
White petrolatum	0.8
Liquid paraffin	8.3
Lanolin	0.8
Stearic acid	16.7
Propylene glycol	3.5
Triethanolamine	1.0
Tween 60	5.0
Methyl paraben	0.1
Water	q.s

Stearic acid amine saponified from triethanolamine and stearic acid acts as a strong oil-in-water emulsifier, and glycerin is a weak water-in-oil emulsifier. Glycerol and ethylparaben are humectant and antiseptic, respectively.

Das, Trailokya et al. formulated a series of herbal cream (O/W type) which was composed of aqueous herbal extract, stearic acid and liquid paraffin in varying concentrations, and they found that formulation F6 (Table 3) satisfied almost all the pharmaceutical parameters tested in this study and presented better issue regeneration and healing capacity [5]. Stearic acid, which occupied 16.7%, acted as an oily phase composition in this formulation.

Usually, cream containing stearic acid could be prepared easily by melting method as reference recording, which can be sketched as melting oily phase and aqueous phase at high temperature separately and then mixing the phases. In detail, the preparation process, as recorded in Richard Lobo et al.'s study, usually contains three steps [41].

1) Preparation of oily phase. Certain amount of stearic acid is taken in a china dish and heated on water bath at 70 °C until 2/3rd of the stearic acid melted. Remaining 1/3rd quantity melted with the waste heat in dish.
2) Preparation of aqueous phase. Aqueous phase substances, such as potassium hydroxide, glycerin and water are also heated on water bath to 70 °C on water bath.
3) Mixing the phases. After melting, both of the phases are mixed by continuous stirring while the air entrapment should be avoided. Preservative is added at the same time. On cooling, perfume and active ingredients are added and triturated well for homogeneous distribution.

Suppository

Suppository is solid-like preparation made from drug and suitable bases for cavity administration. When inserted into cavities, they melt or soften, after which drug dissolution and absorption result in local or systemic effect. Whin rectal suppository is a common form which is usually used as laxatives. Administrated by rectum mucosa, suppositories possess many advantages compared with oral route. Firstly, it is a convenient administration way for infants or patients suffering from vomiting or gastrointestinal disturbances.

What is more, drugs, sensitive to pH conditions and gastrointestinal enzymes or irritating to stomach, could be protected well by absorbed through rectum mucosa. Last but not least, the hepatic first-pass effect can be avoided as the portal circulation is circumvented, contributing to the bioavailability increase of drugs.

The matrices of suppositories are not only important for drug formulation, but also vital to the local or systemic effects of drugs.

The suppository matrices should have certain hardness at room temperature so as to avoid deformation when inserted into cavity.

Meanwhile, they should be softened, melted or dissolved easily so that the medicaments loaded by suppositories can be released after administration. Conventional matrices of suppositories can be classified into hydrophobic and hydrophilic bases.

Stearic acid and its propylene glycol ester are important components in suppository preparation. Propylene glycol stearate is a hydrophobic base for suppository, which is a mixture of propylene glycol monoesters and diesters. The milky white or slightly yellow waxy solid is insoluble in water and inflatable in hot water. With melting point of 36 - 38 °C, propylene glycol stearate is a suitable suppository matrix with high safety and low irritation [42]. Stearic acid is also a main composition in suppository. Sal fat and cocoa butter, in which stearic acid predominates, are another two bases for suppository [6]. Sal fat contains 9,10 dihydroxystearic acid and its mixed triglycerides. The triglycerides contains 9,10 dihydroxy stearic acid 30.5%, stearic acid 57.5%, palmitic acid 6.0% and arachidic acid 5.8%.

Mishra MU et al. prepared rectal suppositories of mesalamine by using Sal fat and evaluated several important *in vitro* characteristics by comparing with the suppositories composed of cocoa butter as standard base. Their findings indicated that Sal fat can be a novel and cost effective suppository base for pharmaceutical formulations.

In order to change the physical properties of suppositories, improve drug absorption and enhance preparation stability, certain additives are needed for suppository preparation, among which hardener is necessary.

Stearic acid, as hardener, is often added into formulation for maintaining hardness of suppositories at room temperature, so as to avoid suppository softening during storage or usage.

Moreover, glycerin monostearate can be used as thickener to adjust drug release speed according to practical clinical requirements.

STEARIC ACID IN ENTERIC COATING

As an endogenous long-chain saturated fatty acid, stearic acid is nontoxic and well biodegradable, and thus used as a common constituent in enteric coating. Preparation with enteric coatings is a kind of oral site-specific drug delivery system. This technique aims at enabling medicaments to pass through the stomach unchanged while disintegrate in a short time span in the fluids of intestines. Two main advantages of drugs delivered in this form can be listed as follows. Firstly, enteric preparations eliminate contact of highly irritating drug with the mucous of stomach and the side effect of gastrointestinal reaction could be avoided consequently. Most importantly, drug inactivation by digestive juice in stomach could be prevented by the enteric coating, which guarantees medicaments to be delivered to the desired portion of intestinal tract in sufficient concentrations and therapeutically active forms.

According to previous study, microspheres coated with stearic acid could reduce the burst release of the encapsulated drugs in acidic stomach, which is of great significance for drugs with short half-life and instability in stomach [7]. It is also found that a highly satisfactory enteric coating could be prepared by using a fatty acid base, such as bases including stearic acid, combined with emulsifying and coating dissolution timing agents [43].

A satisfactory formulation of the enteric coating is shown in table 4. In this formulation, stearic acid, hydrogenated castor oil and castor oil constitute the enteric coating base of proper consistency, and the cholesterol and sodium taurocholate are the emulsifying and timing agents.

This coating can protect the drugs in core from being digested by gastric juice successfully for more than 10 hours, and the emulsifying and timing agents can cuase coating disintegration and drug release when the medicament leaves the stomach and enters the intestines.

A classical enteric capsule preparation can be made by the following steps.

Table 4. The formulation of an enteric coating ([43])

Compositions	Percent (%)
Stearic acid	68
Hydrogenated castor oil	25
Sodium taurocholate	4
Castor oil	2
Cholesterol	1

Firstly, the capsules filled with drugs are emersed in 6.2% alcohol solution of formaldehyde for 10 min at room temperature, so as to protect the capsule from dissolving in stomach. Then, the obtained capsules are soaked in 5% alcohol solution of stearic acid for 5 min at 45 °C, resulting in enteric coating with suitable hardness and friability. Finally, the capsules are put into 16% alcohol solution of phenyl salicytate for 3 min at 40 °C and 20% alcohol solution of paraffin wax for 3 min at 20-25 °C successively. After dried in lime drier for 24 h, the desired enteric capsules are ready.

In addition, stearic acid in enteric coating plays other roles. In Kanazawa et al.'s bubbling enteric coating preparations, stearic acid, as an endogenous fatty acid, served as a safe absorption enhancer to improve the permeability of a pharmaceutically active substance through the biomembranes in digestive tract [44]. Besides, stearic acid could also be used as plasticizer in enteric coatings [45, 46].

STEARIC ACID – A BINDER AGENT

Stearic acid could also be used as a binder agent for pharmaceutical preparation. Binder agent is an auxiliary substance necessary in the process of pelletization before tabletting or capsulation. It is able to make the materials with no or less viscosity gather to be particles or be compressed into molding. Stearic acid is often used together with other binder agents. In 1982, Musikabhumma et al. reported their evaluation of stearic acid and polyethylene glycol as binders for tableting potassium phenethicillin [47]. It is reported that stearic acid melts with a peak temperature of about 58 °C [48] and it has been successfully used in melt pelletization by high shear mixer [49]. With stearic acid as meltable binder, researchers prepared many sustained release drug delivery systems by the single step melt pelletization [8, 48]. In the preparation of paracetamol/stearic acid sustained release delivery system, the composition of the mixtures was paracetamol /lactose /stearic acid: 60 /20 /20, w/w. Paracetamol and lactose were first mixed and heated to 55 °C, after which the stearic acid was added and re-mixed to obtain a uniform distribution of the binder. At this time, the stearic acid reached a molten state (~65 °C). After the subsequent processing, the pellets with homogeneous composition and prolonged drug release behavior were obtained.

In addition, as the binder agent, stearic acid is not only used in pharmaceutics but also in ceramic industry [50, 51].

Stearic Acid – A Traditional Auxiliary in New Drug Delivery System

Although stearic acid is a traditional pharmaceutical auxiliary, we can also find its place in new drug delivery system, like microparticle dispersion system, which has been extensively studied in recent decades.

General examples of this system are polymer micelles, microsphere, nanoparticles, liposomes, etc. The microparticle dispersion system has the following advantages:

1 Small size and high dispersion degree contribute to the enhanced solubility, dissolution rate and bioavailability;
2 Being encapsulated into vehicles, the stability of drugs can be enhanced either *in vitro* or *in vivo*;
3 The *in vivo* distribution of particles in different sizes appeared to be selective, which is termed as passive targeting;
4 Based on the characteristics of microparticles, packaged drugs often exhibit longer action time and lower side effect *in vivo*.

The carrier materials of many microparticle dispersion systems are usually synthesized polymers whose accumulation in vivo could cause possible toxicity. In contrast, stearic acid is a better material for microparticle dispersion system due to its biocompatibility and nontoxicity. Generally, as solid lipid with the above advantages, stearic acid can be used for preparation of solid lipid nanopartical (SLN), which is supposed to be more stable than emulsion or liposome due to solid form of stearic acid at room temperature.

Qiang Zhang et al. prepared stearic acid nanoparticles loaded with cyclosporin A by the classical and simple melt - homogenization method. The mean size and encapsulation efficiency of the cyclosporin A stearic acid nanoparticles are 316.1 nm and 88.36%, respectively. In addition, no chemical reaction between cyclosporin A and stearic acid was observed, and the nanoparticles exhibited improved bioavailability and sustained drug release property [9]. Woo JO et al. improved this technique to prepare salicylic acid loaded nanoparticles, wherein they combined stearic acid and oleic acid in the lipid phase and prepared the nanoparticles by melt emulsification method and further ultrasonication technique. In this formulation, different amount of oleic acid contributed to different degrees of crystallinity of the nanoparticles matrix, thereby improving the encapsulation efficiency.

The optimized stearic acid - oleic acid nanoparticles were subsequently incorporated in cream and showed sustained *in vitro* release feature [10]. Moreover, the size and drug release rate of stearic acid nanoparticles could be changed by adding different amount of oleic acid in the formulation [10, 11].

Robson H et al. prepared the cefuroxime axetil stearic acid microspheres to cover the unpalatable drug, in which the stearic acid coated drug particles via a spray chilling process, and they also investigated deeply the drug release properties of the microsphere in different buffer media [52, 53].

These particles are subsequently used in the form of suspension which shows acceptable flavor and taste and suitable bioavailability in children [54].

CONCLUSION

Based on the above description, it is no doubt to say that stearic acid along with its derivatives are important and promising materials in biomedical science. With the progress of chemical industry, stearic acid of high purity and quality, or various stearic acids derivatives can be produced to meet different requirements of medical application.

Nowadays, nontoxic and biodegradable materials have been attracting increasing attention of researchers in biomedical science.

With the development of biomedicine and biomaterial, more and more properties of stearic acids will be found and applied in biomedicine for improving the living and health conditions of human beings.

REFERENCES

[1] Phadke, D. S., Keeney, M. P., Norris, D. A. Evaluation of batch-to-batch and manufacturer-to-manufacturer variability in the physical properties of talc and stearic acid. *Drug Dev. Ind. Pharm.* 1994;20(5):859-871.

[2] Strickley, R. G. Solubilizing excipients in oral and injectable formulations. *Pharm. Res.* 2004;21(2):201-230.

[3] Bárány, E., Lindberg, M., Lodén, M. Unexpected skin barrier influence from nonionic emulsifiers. *Int. J. Pharm.* 2000;195(1):189-195.

[4] Peper, H. The defoaming of synthetic detergent solutions by soaps and fatty acids. *J. Colloid Science.* 1958;13(3):199-207.

[5] Das, T., Debnath, J., Nath, B., Dash, S. Formulation and evaluation of an herbal cream for wound healing activity. *Int. J. Pharm. Pharm. Sci.* 2014;6(2):693-697.

[6] Mishra, M. U., Maskare, R. G. Preliminary evaluation of mesalamine suppositories using sal fat as novel base. *J. Adv. Sci. Res.* 2013;4(3):37-40.

[7] Angadi, S. C., Manjeshwar, L. S., Aminabhavi, T. M. Stearic acid-coated chitosan-based interpenetrating polymer network microspheres: Controlled release characteristics. *Ind. Eng. Chem. Res.* 2011;50(8): 4504-4514.

[8] Thomsen, L. J., Schaefer, T., Kristensen, H. Prolonged release matrix pellets prepared by melt pelletization II. Hydrophobic substances as meltable binders. *Drug Dev. Ind. Pharm.* 1994;20(7):1179-1197.

[9] Zhang, Q., Yie, G., Li, Y., Yang, Q., Nagai, T. Studies on the cyclosporin A loaded stearic acid nanoparticles. *Int. J. Pharm.* 2000;200 (2):153-159.

[10] Woo, J. O., Misran, M., Lee, P. F., Tan, L. P. Development of a controlled release of salicylic acid loaded stearic acid-oleic acid nanoparticles in cream for topical delivery. *The Scientific World J.* 2014; 2014:205703.

[11] Hu, F. Q., Jiang, S. P., Du, Y. Z., Yuan, H., Ye, Y. Q., Zeng, S. Preparation and characterization of stearic acid nanostructured lipid carriers by solvent diffusion method in an aqueous system. *Colloids Surfaces B Biointerfaces.* 2005;45(3):167-173.

[12] Ducatman, F. P., Flanagan, J. D. Stable sodium aspirin tablet compositions. *US patent 4,686,212.* Aug. 11, 1987.

[13] Rowe, R. The adhesion of film coatings to tablet surfaces-the effect of some direct compression excipients and lubricants. *J. Pharm. Pharmacol.* 1977;29(1):723-726.

[14] Levy, G., Gumtow, R. H. Effect of certain tablet formulation factors on dissolution rate of the active ingredient III. Tablet lubricants. *J. Pharm. Sci.* 1963;52(12):1139-1144.

[15] Iranloye, T. A., Parrott, E. L. Effects of compression force, particle size, and lubricants on dissolution rate. *J. Pharm. Sci.* 1978;67(4):535-539.

[16] Shibata, D., Shimada, Y., Yonezawa, Y., Sunada, H., Otomo, N., Kasahara, K. Application and evaluation of sucrose fatty acid esters as lubricants in the production of pharmaceuticals. *J. Pharma. Sci. Technol.* 2002;62(4):133-145.

[17] Billany, M., Richards, J. Batch variation of magnesium stearate and its effect on the dissolution rate of salicylic acid from solid dosage forms. *Drug Dev. Ind. Pharm.* 1982;8(4):497-511.

[18] Lehtola, V., Heinämäki, J., Nikupaavo, P., Yliruusi, J. Effect of Some Excipjents and Compression Pressure on the Adhesion of Aqueous-Based Hydroxypropyl Methylcellulose Film Coatings to Tablet Surface. *Drug Dev. Ind. Pharm.* 1995;21(12):1365-1375.

[19] Late, S. G., Yu, Y.-Y., Banga, A. K. Effects of disintegration-promoting agent, lubricants and moisture treatment on optimized fast disintegrating tablets. *Int. J. Pharm.* 2009;365(1-2):4-11.

[20] Mitrevej, K. T., Augsburger, L. Adhesion of tablets in a rotary tablet press II. Effects of blending time, running time, and lubricant concentration. *Drug Dev. Ind. Pharm.* 1982;8(2):237-282.

[21] Bangudu, A., Pilpel, N. Effects of composition, moisture and stearic acid on the plasto-elasticity and tableting of paracetamol-microcrystalline cellulose mixtures. *J. Pharm. Pharmacol.* 1985;37(5):289-293.

[22] Myers, D. *Surfactant science and technology.* 2005, pp. 6 (in English).

[23] Myers, D. *Surfactant science and technology.* 2005, pp. 313 (in English).

[24] Schmidts, T., Dobler, D., Nissing, C., Runkel, F. Influence of hydrophilic surfactants on the properties of multiple W/O/W emulsions. *J. Colloid Interface Sci.* 2009;338(1):184-192.

[25] Macierzanka, A., Szelag, H., Moschakis, T., Murray, B. S. Phase transitions and microstructure of emulsion systems prepared with acylglycerols/zinc stearate emulsifier. *Langmuir.* 2006;22(6):2487-2497.

[26] Fitzhugh, O. G., Bourke, A. R., Nelson, A. A., Frawley, J. P. Chronic oral toxicities of four stearic acid emulsifiers. *Toxicol. Appl. Pharm.* 1959;1(3):315-331.

[27] Myers, D. *Surfactant science and technology.* 2005, pp. 119 (in English).

[28] Myers, D. *Surfactant science and technology.* 2005, pp. 6, 194 (in English).

[29] Lavasanifar, A., Samuel, J., Sattari, S., Kwon, G. S. Block copolymer micelles for the encapsulation and delivery of amphotericin B. *Pharm. Res.* 2002;19(4):418-422.

[30] Yan, J., Du, Y. Z., Chen, F. Y., You, J., Yuan, H., Hu, F. Q. Effect of proteins with different isoelectric points on the gene transfection efficiency mediated by stearic acid grafted chitosan oligosaccharide micelles. *Mol. Pharm.* 2013;10(7):2568-2577.

[31] Muzíková, J., Vyhlídalová, B., Pekárek, T. A study of micronized poloxamers as lubricants in direct compression of tablets. *Acta Pol. Pharm.* 2012;70(6):1087-1096.
[32] Du, Y. Z., Lu, P., Yuan, H., Zhou, J. P., Hu, F. Q. Quaternary complexes composed of plasmid DNA/protamine/fish sperm DNA/stearic acid grafted chitosan oligosaccharide micelles for gene delivery. *Int. J. Biol. Macromol.* 2011;48(1):153-159.
[33] Zhou, Y. Y., Du, Y. Z., Wang, L., Yuan, H., Zhou, J. P., Hu, F. Q. Preparation and pharmacodynamics of stearic acid and poly (lactic-co-glycolic acid) grafted chitosan oligosaccharide micelles for 10-hydroxycamptothecin. *Int. J. Pharm.* 2010;393(1-2):143-151.
[34] Ye, Y. Q., Yang, F. L., Hu, F. Q., Du, Y. Z., Yuan, H., Yu, H. Y. Core-modified chitosan-based polymeric micelles for controlled release of doxorubicin. *Int. J. Pharm.* 2008;352(1):294-301.
[35] You, J., Hu, F. Q., Du, Y. Z., Yuan, H. Polymeric micelles with glycolipid-like structure and multiple hydrophobic domains for mediating molecular target delivery of paclitaxel. *Biomacromolecules.* 2007;8(8):2450-2456.
[36] Hu, P., Wang, T., Xu, Q., et al. Genotoxicity evaluation of stearic acid grafted chitosan oligosaccharide nanomicelles. *Mutat. Res.* 2013;751 (2): 116-126.
[37] Hans, G. Defoaming agents. *US patent 2,753,309.* July 3, 1956.
[38] Lamont, W. A. Fatty acid esters of polyoxypropylated glycerol. *US patent 3,337,595.* Aug. 22, 1967.
[39] Shi, M. L., Wang, D. G., Wang, H. Y., Sun, X. Y. Relationships between antifoaming potential and the number of hydrophile groups as well as HLB of polyoxyalkylene ethers and its stearates. *Chem. J. Chinese Universities.* 1988;9(6):640-642 (in Chinese).
[40] Xiao, S. G. *New handbook of National Standards for Pharmaceutical Excipients.* 2006, pp. 914 (in Chinese).
[41] Lobo, R., Prabhu, K., Shirwaikar, A., Shirwaikar, A., Ballal, M. Formulation and Evaluation of Antiseptic Activity of the Herbal Cream Containing Curcuma longa and Tea Tree Oil. *J. Biol. Active Prod. Nat.* 2011;1(1):27-32.
[42] Schlueter, K. W., Schulz-Kaiser, E. Process and composition for the production of suppositories. *US patent 4,151,274.* Apr. 24, 1979.
[43] Kuever, R. A., Maney, P. V. Enteric coating. *US patent 2,373,763.* Apr. 17, 1945.

[44] Kanazawa, H., Shimizu, K., Sasaki, K., Sugimoto, T. Bubbling enteric coated preparations. *US patent 6,326,360 B1.* Dec. 4, 2001.

[45] Behl, C., Beskid, G., Shah, N., Tossounian, J., Unowsky, J. Enteric coated oral dosage form. *US patent 4,525,339.* Jun. 25, 1985.

[46] Liu, F., Merchant, H. A., Kulkarni, R. P., Alkademi, M., Basit, A. W. Evolution of a physiological pH6. 8 bicarbonate buffer system: Application to the dissolution testing of enteric coated products. *Eur. J. Pharm. Biopharm.* 2011;78(1):151-157.

[47] Musikabhumma, P., Rubinstein, M. H., KA K. Evaluation of stearic acid and polyethylene glycol as binders for tabletting potassium phenethicillin. *Drug Dev. Ind. Pharm.* 1982;8:169-188.

[48] Grassi, M., Voinovich, D., Moneghini, M., Franceschinis, E., Perissutti, B., Filipovic-Grcic, J. Preparation and evaluation of a melt pelletised paracetamol/stearic acid sustained release delivery system. *J. Control Release* 2003;88(3):381-391.

[49] Voinovich, D., Moneghini, M., Perissutti, B., Franceschinis, E. Melt pelletization in high shear mixer using a hydrophobic melt binder: influence of some apparatus and process variables. *Eur. J. Pharm. Biopharm.* 2001;52(3):305-313.

[50] Lin, S., German, R. Interaction between binder and powder in injection moulding of alumina. *J. Mater. Sci.* 1994;29(19):5207-5212.

[51] Jaw, K.-S., Hsu, C.-K., Lee, J.-S. The thermal decomposition behaviors of stearic acid, paraffin wax and polyvinyl butyral. *Thermochim. acta.* 2001;367:165-168.

[52] Robson, H., Craig, D., Deutsch, D. An investigation into the release of cefuroxime axetil from taste-masked stearic acid microspheres. III. The use of DSC and HSDSC as means of characterising the interaction of the microspheres with buffered media. *Int. J. Pharm.* 2000;201(2):211-219.

[53] Robson, H., Craig, D., Deutsch, D. An investigation into the release of cefuroxime axetil from taste-masked stearic acid microspheres. II. The effects of buffer composition on drug release. *Int. J. Pharm.* 2000; 195 (1):137-145.

[54] Shalit, I., Dagan, R., Engelhard, D., Ephros, M., Cuningham, K. Cefuroxime efficacy in pneumonia: sequential short-course i.v./oral suspension therapy. *Isr. J. Med. Sci.* 1994;30(9):684-689.

In: Stearic Acid
Editors: Yunfeng Lin and Qiang Peng
ISBN: 978-1-63463-172-3

Chapter 4

LUBRICANT POTENTIAL OF STEARIC ACID AND DERIVATIVES FOR THE PRODUCTION TABLETS BY DIRECT COMPRESSION

John Rojas*
Department of Pharmacy, School of Pharmaceutical Chemistry, University of Antioquia, Colombia

ABSTRACT

Stearic acid is a saturated, waxy, fatty acid, which is extracted from animal or vegetable fats and oils. In nature, stearic acid occurs primarily as a mixed triglyceride, or fat, with other long-chain acids and as an ester of a fatty alcohol (i.e., coconut oil and cocoa butter). It is much more abundant in animal fat than in vegetable fat, lard and tallow. Currently, stearic acid is derived mostly from palm oil and contains no trans-fatty acids. Stearic acid is generally regarded as safe (GRAS) and can be added in small amounts (usually <5%) to pharmaceutical products.

Stearic acid and its derivatives such as magnesium stearate and sodium stearyl fumarate are used as the lubricants of choice for the production of solid dosage forms. They improve density, stickiness, and flow to powdered excipient-drug mixtures. Further, they are used as a binder that helps tablets hold together and break apart properly. In this chapter, the effect of stearic acid, magnesium stearate and sodium stearyl fumarate on the particle and tableting properties of model excipients such

* Corresponding author: Email: jrojasca@gmail.com (JR).

as lactose monohydrate and sorbitol was assessed by producing compacts using the direct compression technology.

The lubricant blending time had a larger effect on the particle and tableting properties of sorbitol than lactose monohydrate. Properties such as flowability, porosity and compact tensile strength depended more on the particle size, surface roughness and shape of the excipient than on the lubricant level. However, the amount of lubricant increased densification and flow, but decreased porosity, compressibility, compactibility and ejection forces. In most cases, powder flowability improved due to a decrease in friction and surface roughness and adhesion among powder particles. Opposed to the sorbitol behavior, the rough surface and irregular shape of the fragmenting lactose monohydrate eased the adherence of lubricant in the cavities and pores, making the formation of a lubricant film irregular and hence, lactose was less susceptible to the lubricant action.

INTRODUCTION

Lubricants are pharmaceutical aids that increase the die filling properties of a drug-excipient mixture mainly due to their ability to improve flow, as well as to minimize die-wall friction and prevent powder adhesion to punch faces. As a result, the tablet weight variability in the tableting process decreases, improving content uniformity, and enhancing the quality of the manufacturing process [1]. Further, lubricants also help minimizing the electrostatic forces in the powder interfaces caused by friction. These forces are different from cohesion forces which make the particles to stick together due to the interparticulate attraction between their surfaces [2].

In general, lubricants can be added internally (boundary lubricants) or externally. In the former case, lubricants are added to the powder mixture and blended briefly to coat the particles, whereas in the latter, lubricants are dissolved in an organic solvent and sprayed onto the surface of the punches to form a film [3]. In this case, only the lower punch and die, not the final blend, are lubricated and thus, the external addition is only useful when tablet properties are very sensitive to lubricants.

Hydrophobic lubricants such as stearic acid, magnesium stearate and sodium stearyl fumarate are more efficient than hydrophilic lubricants such as talc, calcium silicate and fumed silica [4]. The amphiphilic property of hydrophobic lubricants, allows them to adhere to metal surfaces where their polar head orientates apart from the carbohydrate tail forming a boundary film on the particles or metal surfaces. This mechanism reduces the contact area and intrinsic adhesive interactions between the powder and the metallic surfaces preventing the formation of tight junctions between the particles and the die-wall.

It is widely accepted that hydrophobic lubricants show no toxic effects since they have a poor dissolution and absorption in the gastrointestinal tract [5]. They are commonly used in the range of 0.25-5.0% (w/w) for tablet compression (Table 1). However, at very low levels could function as a glidant (flow enhancer). The use of an excessive amount of lubricant could impede flow because the interparticulate adhesion forces would surpass the interparticulate friction forces between the host particles. Moreover, lubricants are always added last after all other components have been thoroughly blended to minimize their deleterious effect. The optimal blending time (usually from 2 to 15 min) depends on their chemistry and amount since it determines the degree particle coating [6].

Stearic acid can be obtained from plant and animal sources. Typicaly, it is obtained by hydrolysis of shea or cocoa butter at high temperatures and pressures, followed by distillation. On the other hand, stearic acid is the main raw material to produce magnesium stearate and sodium stearyl fumarate. Therefore, magnesium stearate is produced by saponification of stearic acid using sodium hydroxide followed by precipitation with magnesium sulfate. Likewise, sodium stearyl fumarate is obtained by esterification of stearic acid with fumaric acid followed by partial neutralization with sodium hydroxide [7]. The aliphatic character of the C_{18} chain provides the water repellency properties (hydrophobicity). Further, the present of a metal such as magnesium or sodium in the molecule increases hydrophobicity and melting point becoming useful in a high range of temperatures and preventing the powder from absorbing water and thus, from forming agglomerates (Figure 1 and Table 1). In other words, the lubricant power increases by the presence of a metal in the aliphatic structure. However, the presence of a metal increases the incompatibility with acid drugs such as ascorbic acid and aspirin [8].

The low melting point of stearic acid makes it ideal as a dispersive agent for the production of soaps, shampoos, detergents and shaving creams. They are added to these products melted and allowed to cool down at room temperature with a gentle stirring to recrystallize and form a semisolid and stable structure. Further, due to its hydrophobic properties and the ability to form hard compacts has been used for controlled release of some drugs.

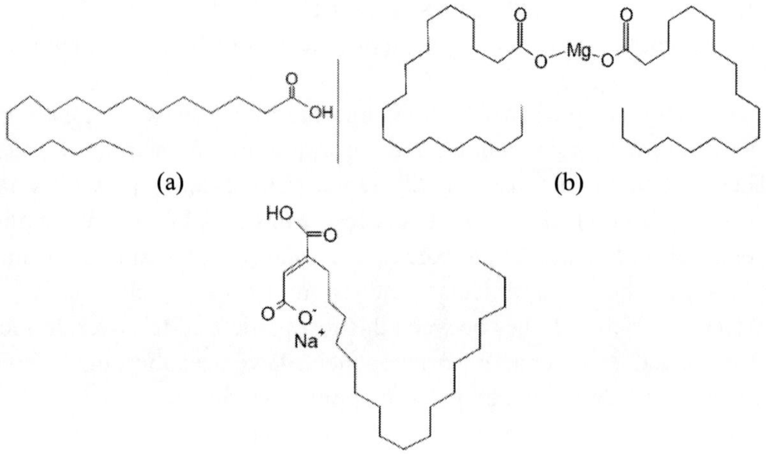

(a) (b)

(c)

Figure 1. Chemical structure of stearic acid and derivatives: (A) Stearic acid, (B) magnesium stearate and (C) sodium stearyl fumarate.

Table 1. Characteristics of stearic acid and derivatives

Property	Stearic acid	Magnesium Stearate	Sodium stearyl fumarate
Molecular weight (g/mol)	284.5	591.3	390.5
Melting point (°C)	49	151	225
Bulk density (g/cm^3)	0.51	0.25	0.17
Moisture content (%)	2.1	2.3	3.2
Water solubility (25 °C)	3 mg/L	40 mg/L	Insoluble
Contact angle (degrees)	81.04	102.3	92.07
Mean particle size (μm)	195.4±5.5	5.45± 0.25	23.2±0.02
Typical tableting level used (%)	1-5	0.2-2	0.5-2

In this chapter, the effect of hydrophobic lubricant (i.e, stearic acid, magnesium stearate and sodium stearyl fumarate), their concentration and blending time on the powder and compaction properties of a plastic deforming (sorbitol) and fragmenting material (lactose monohydrate) alone and in mixtures with spironolactone was assessed.

PHYSICOCHEMICAL CHARACTERIZATION

Fourier Transform Infrared (FT-IR) Characterization

This type of spectroscopy represents a fingerprint of a material with absorption bands, which correspond to the frequencies of vibrations between the bonds of the atoms forming the material. Due to the different combination of atoms, each material has a unique infrared spectrum and hence, this technique is useful for qualitative and quantitative analyses [9]. The FT-IR spectra of the lubricants are shown in Figure 2. Stearic acid and derivatives showed vibrational bands at 2918 and 2851 cm^{-1} assigned to the asymmetric and symmetric stretching vibrations, respectively for the methylene groups (CH_2) of the aliphatic chain. The band at 1704 cm^{-1} (1719 cm^{-1} for sodium stearyl fumarate) is assigned to the carbonyl stretching and was absent for magnesium stearate, which in turn presented a band at 1586 cm^{-1}. Further, the band at 1463-1467 cm^{-1} is attributed to the deformation of methylene groups and the band at 1297-1315 cm^{-1} is assigned to the symmetric deformation of the methyl group [10].

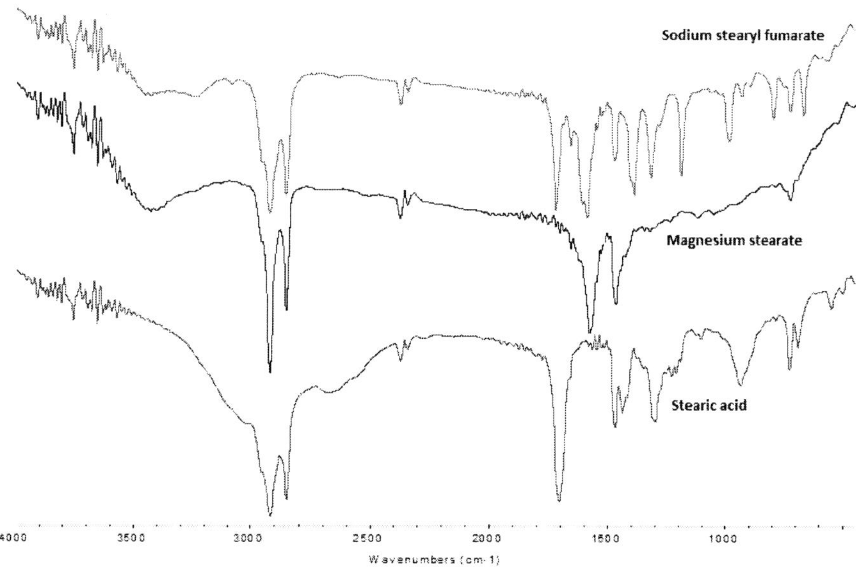

Figure 2. FT-IR spectra of stearic acid and derivatives.

Powder X-Ray (PXRD) Characterization

This technique measures the degree of interference pattern of X-rays waves when they bombard the atoms of a crystal. The diffraction pattern is related to the structure of the crystal. Powder X-Ray diffractions were obtained on a PANalytical diffractometer (Model, Empyrean 2012, Westborough, MA) at 45 kV and 40 mA, equipped with a monochromatic CuK_α (α_1 =1.540598 Å, α_2 =1.544426 Å) X-Ray radiation. Diffractograms were obtained over a 5 to 45° range at a scan step and step time of 0.039 and 38.2 s, respectively.

Figure 3 shows the diffractograms of stearic acid and derivatives. In the long spacing region of stearic acid diffractogram (low values of 2θ angles), there were four main diffraction peaks at 7.1, 11.7, 21.5 and 23.9 degrees which corresponded to the typical reflections for the E allomorph. Likewise, magnesium stearate exists as an anhydrous, monohydrate, dihydrate and trihydrate forms. In this case, the difractogram showed the main diffraction peaks at 5.5, 9.1, 21.8, 22.4 and 23.3° 2θ, which corresponded to the reflections of the trihydrate form. On the other hand, sodium stearyl fumarate showed a series of diffraction peaks at 7.2, 8.6, 11.4, 12.9, 15.7, 20.1, 21.9 and 40.2° 2θ, where the peak at 8.6 was the most intense observed [11].

Functional Properties on Powder Blends

Effect of Blending Time on the Lubricant Power

Batches of ~20 g of a lubricant:filler (99:1 ratio) were passed freely through a number 30 mesh sieve (600 μm) and then blended separately in a V-blender (Riddhi Pharma Machinery, Gulabnagar, India) at 5, 10 and 15 minutes. Lactose monohydrate and sorbitol were used as model fragmenting and plastic deforming excipients, respectively.

The true density (ρ_{true}) was measured on approximately 2 g of sample employing a Helium pycnometer (AccupycII 1340, Micromeritics, USA). Powder porosity, bulk and tap densities were determined as reported previously [12]. Flow rate was determined on ~20 g of powder by recording the time taken to pass freely through a glass funnel (13.1 mm diameter). Moisture content was determined on an Ohaus moisture analyzer (MB 200, NJ, USA) loaded with ~3g of sample and operated at 100°C for 5 min. The mean particle size of lubricants was determined by optical microscopy (700X magnification) using an optical microscope (BM-180, Boeco, Germany) coupled with a digital camera (DCR-SX45, Sony, Tokyo, Japan) as reported previously [9]. Likewise, powder compressibility and the particle size distribution of the excipient-spironolactone mixtures were obtained on 20 g of sample employing an AutoTap® and RoTap® sieve shaker for 10 minutes as reported previously [13].

The effect of blending time and hence, the extent of a lubricant coating on the excipient properties is listed on Table 2. Powder flow depended more on the excipient, rather than the lubricant chemistry. For instance, the flow rate of mixtures containing sorbitol was 55 times larger than those of containing lactose independent of the blending time. This is attributed to the large and more regular particle size of sorbitol, which in turn, was highly influenced by gravity and hence, could flow into the dies faster than samples containing lactose as filler. On the other hand, large blending times led to an increase in compressibility and powder porosity. This could be explained by the coating powder of lubricants, which was more effective when large blending times were employed. Interestingly, blends containing sorbitol showed more lubricant susceptibility since they were less compressible and presented less void spaces than lactose mixtures.

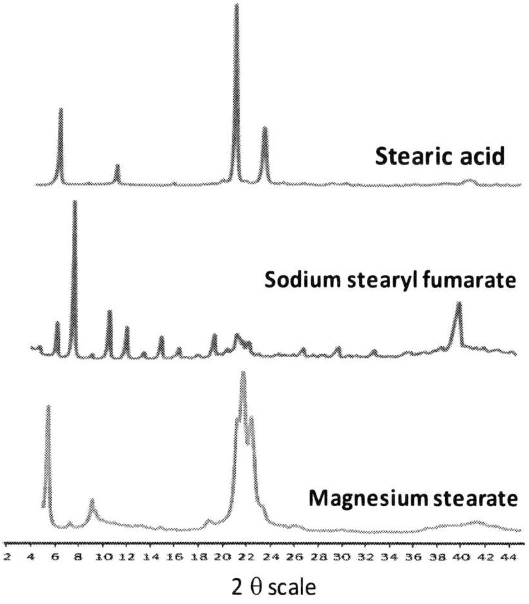

Figure 3. Powder X-Rays difractograms of stearic acid and derivatives.

Table 2. Effect of blending time on the powder and tableting properties of lubricant-excipient mixtures

Filler	Lubricant	Blending time (min)	Flow rate (g/s)	Compressibility (%)	Powder porosity (%)	Compact tensile strength (MPa)
Lactose monohydrate	Stearic acid	5	0.1	34.4	67.1	0.3
		10	0.1	36.6	69.5	0.2
		15	0.1	37.2	69.5	0.2
	Magnesium stearate	5	0.1	43.1	70.0	0.2
		10	0.4	42.4	70.0	0.2
		15	0.3	40.2	68.7	0.3
	Sodium stearyl fumarate	5	0.5	38.3	70.2	0.3
		10	0.4	40.8	73.5	0.3
		15	0.3	39.6	73.7	0.2
Sorbitol	Stearic acid	5	8.7	16.3	63.7	3.8
		10	7.0	19.4	64.4	3.6
		15	8.4	18.1	63.7	2.8
	Magnesium stearate	5	15.3	8.0	51.1	2.9
		10	16.5	11.3	51.1	3.0
		15	15.5	15.5	63.7	2.9
	Sodium stearyl fumarate	5	6.9	19.5	59.9	3.5
		10	7.1	22.7	65.1	3.4
		15	6.5	28.5	65.8	3.3

Compactibility is the ability of a material to form a compact with sufficient strength when a compression force is applied. A ductile binder can withstand large deformations without breaking, whereas a brittle material fails having a negligible deformation. Both, ductility and brittleness favor bonding because new surfaces are produced and consequently, an increase in contact area between particles occurs at compression.

Compactibility, as expressed by the compact tensile strength of lactose (brittle deforming material) was lower than the one of sorbitol (plastic deforming material). Therefore, lactose was virtually not affected by blending time since new lubricant-free surfaces are constantly formed counteracting the effect of lubricants. On the contrary, sorbitol had an effective particle coating and thus, the formation of bonding points between particles required for the formation of strong compacts was more restricted and thus, compactibility decreased with extended blending times.

Effect of the Lubricant Load

In order to determine the effect of the lubricant level and excipient plasticity, a new set of batches were produced at lubricant levels of 1.0, 1.5 and 2 %. The batch size and blending time were 20 g and 15 minutes, respectively. In this case, a lactose:sorbitol at 1:0, 1:1 and 0:1 ratios were also included in the blend to determine the prevalence of a deformation mechanism.

The particle size slightly decreased or increased when treated with lubricants for lactose and sorbitol, respectively (Table 3). The former mechanism is explained by the lubricant action, which favors the passage of particles through the sieves of smaller sizes. On the contrary, for materials of a large size such as sorbitol an aggregate particle structure is formed blocking the particle migration through the sieves. Likewise, the interfacial lubricant action was responsible for the increase in densification, packing and rearrangement of the particles in the powder bed easing the compression process of the particles when an external force was applied. As a result, the bulk and tap densities are increased by the lubricant action. In the same way, the high compression ability of the powder bed forced the particles to have a low porosity and hence, a high packing tendency.

Powder blends with a high porosity such as those containing lactose monohydrate had more void spaces (low bulk density) to accommodate a lubricant, resulting in a low lubricant sensitivity. Conversely, sorbitol blends showed a large sensitivity to lubricants. Furthermore, high amounts of lubricants interfered with bonding forces between particles and softened tablets. The magnitude of the lubricant sensitivity depended on the excipient deformation mechanism and lubricant level. This sensitivity was larger in ductile than in brittle materials. Brittle materials (i.e., lactose monohydrate) fracture and fragment during compaction. Therefore, as more fresh surfaces not coated by lubricant particles are generated, they tend to bond together. On the other hand, the continuous lubricant film on ductile (i.e., sorbitol) particles is not destroyed weakening particle bonding as particles with less fresh surfaces are available restricting the free crystal-plane slipping.

Table 3. Effect of lubricant level on the powder and tableting properties of lubricant-excipient mixtures

Filler	Lubricant	Level (%)	Mean diameter (μm)	Flow rate (g/s)	Tap density (g/cm^3)	Bulk density (g/cm^3)	Moisture content (%)	Porosity (%)	Compressibility (%)	Lubricant sensitivity ratio	Ejection force (MPa)
Lactose Monohydrate	Stearic acid	1.0	146.3	0.08	0.75	0.48	4.5	69.5	37.2	0.11	4.3
		1.5	109.9	0.07	0.72	0.50	4.5	67.7	31.2	0.14	2.8
		2.0	102.2	0.06	0.71	0.49	0.4	67.7	32.9	0.18	2.3
	Magnesium stearate	1.0	107.4	0.01	0.80	0.49	3.4	68.6	40.2	0.25	3.2
		1.5	101.1	2.14	0.80	0.56	1.4	64.0	31.0	0.32	2.8
		2.0	102.4	3.75	0.90	0.88	3.0	43.6	16.2	0.36	2.0
	Sodium stearyl fumarate	1.0	104.6	0.29	0.67	0.41	4.1	73.7	39.5	0.18	3.1
		1.5	102.4	0.42	0.72	0.48	4.3	69.5	35.1	0.14	1.9
		2.0	96.7	0.08	0.71	0.50	3.0	68.1	31.1	0.18	1.9
Sorbitol	Stearic acid	1.0	294.8	8.38	0.62	0.54	3.8	64.1	18.0	0.46	3.1
		1.5	320.1	8.57	0.59	0.51	2.9	66.2	14.1	0.50	2.3
		2.0	351.3	9.43	0.65	0.55	3.7	63.5	14.4	0.51	2.3
	Magnesium stearate	1.0	305.4	3.88	0.65	0.55	4.0	64.1	15.5	0.51	2.1
		1.5	318.2	12.90	0.66	0.58	3.8	61.8	11.8	0.52	2.1
		2.0	330.6	9.43	0.90	0.97	3.9	36.1	13.9	0.52	2.0
	Sodium stearyl fumarate	1.0	316.3	6.49	0.67	0.52	4.6	65.8	22.5	0.46	2.0
		1.5	336.0	7.65	0.69	0.59	4.9	66.8	13.7	0.51	1.8
		2.0	352.9	6.72	0.80	0.88	4.7	42.3	16.8	0.59	1.7

TABLETING PROPERTIES

Experimental Design and Statistical Analysis

In order to determine the effect of three factors named as lubricant hydrophobicity (given by their contact angle), lubricant level (%) and excipient brittleness (given by the respective yield pressure value) a Box Behnken experimental design was employed and the levels of each factor are shown in Table 4. The batch size, spironolactone (hydrophobic drug) and sodium starch glycolate (disintegrant) levels remained constant at 20 g, 4 g and 0.4 g, respectively. The Design Expert® software (vs. 9.02, StatEase, Inc, Minneapolis, MN) was employed for the statistical analysis. The responses tested corresponded to the amount of drug dissolve after 60 minutes, compact disintegration time and particle size.

Compaction Properties

Cylindrical compacts of ~500 mg were manufactured on an instrumented single station tablet press (Compac 060804, Indemec, Columbia) equipped with a 13-mm flat-faced punches and die tooling at a dwell time of 1 s and sufficient compression pressure to achieve ~20% porosity. The compact ejection forces were measured directly from a load cell (LCGD-10K, Omega Engineering, Inc, Stanford, CT). The resulting compacts were tested for tensile strength, lubricant sensitivity and disintegration time as reported previously [13, 14].

Table 4. Factors and levels of the Box Behnken experimental design

Factor	Level		
	Low	Medium	High
Hydrophobicity (degrees)	81.0 (stearic acid)	92.1 (sodium stearyl fumarate)	102.3 (magnesium stearate)
Lubricant level (%)	1.0	1.5	2.0
Excipient brittleness (MPa^{-1})	21.2 (sorbitol)	72.5 (sorbitol:lactose, 1:1)	123.7 (lactose monohydrate)

Table 5. Box Behnken experimental matrix

Batch	Hydrophobicity	Level	Brittleness	Drug released (%)	Disintegration time (s)	Particle Size (µm)
1	-1 (81.0)	0 (1.5)	1 (123.7)	69.8	32.3	106.6
2	-1 (81.0)	1 (2.0)	0 (72.5)	72.6	458.3	149.5
3	0 (92.1)	0 (1.5)	0 (72.5)	70.8	441.3	149.5
4	0 (92.1)	0 (1.5)	0 (72.5)	49.5	396.0	149.5
5	1 (102.3)	-1 (1.0)	0 (72.5)	70.2	409.3	149.5
6	1 (102.3)	0 (1.5)	1 (123.7)	59.0	161.7	98.4
7	0 (92.1)	1 (2.0)	1 (123.7)	64.4	33.0	102.2
8	0 (92.1)	1 (2.0)	-1 (21.2)	68.5	485.7	209.3
9	1 (102.3)	1 (2.0)	0 (72.5)	64.0	502.0	149.5
10	1 (102.3)	0 (1.5)	-1 (21.2)	37.0	635.3	210.3
11	0 (92.1)	0 (1.5)	0 (72.5)	41.6	512.7	155.3
12	0 (92.1)	-1 (1.0)	1 (123.7)	63.5	53.3	97.9
13	0 (92.1)	-1 (1.0)	-1 (21.2)	58.0	501.0	210.3
14	-1 (81.0)	-1 (1.0)	0 (72.5)	47.4	600.7	148.3
15	-1 (81.0)	0 (1.5)	-1 (21.2)	67.2	346.3	209.8

Compaction studies showed that during tablet compression, an ejection force was needed to overcome the residual die-wall force to push the tablet out of the die. This ejection force occurs by the interparticulate interactions of the powder blend and measures the adhesive interactions between the tablet and die-wall surface. Therefore, during tablet ejection the bond between the tablet sides and the die-wall is broken up by pushing the tablet out of the die (scraping action). Interestingly, the ejection force decreased with increasing levels of lubricants. For instance, lactose blends required higher compression and ejection forces than those containing sorbitol. In fact, the ejection forces of lactose blends were even higher than the actual compact tensile strength indicating the formation of high friction forces between the die-wall and the tablet. For this reason, the compression process for lactose mixtures was more complex than sorbitol.

Lubricant Effect on Spironolactone Dissolution and Compact Disintegration

The concentration of spironolactone (uniformity of the dose) in each powder mixture was found by UV analysis (HACH DR500, HACXH Company, Loveland, CO) at 242 nm. A calibration curve was built at 2.5, 5, 10, 20 and 30 µg/mL concentrations. Dissolution studies were conducted employing an Erweka (DT6-K, Erweka GmbH, Milford, CT) type 2 apparatus containing 1L of 0.1N HCl and 0.1 % of sodium lauryl sulfate at 37°C and 75 rpm for 1h. At 5, 15, 30, 45 and 60 min, aliquots of 1mL each, were taken, filtrated and diluted (200 µL/1 mL) before measurement.

In general, lubricants had a negative effect on the *in-vitro* dissolution of spironolactone tablets, with the stearic acid derivatives having a more pronounced deleterious effect (Table 5). Thus, lubricants greatly increased hydrophobicity (the resistance of a material to absorb water), decreasing water penetration rate and hence, hindering drug dissolution and tablet disintegration. For this reason, only blends containing lactose released more than 70% of spironolactone within 60 min.

Compact disintegration of materials having a poor compactibility (i.e., lactose) was faster than that shown by sorbitol, which formed harder and slow dissolving compacts.

A Box Behnken experimental design was used to predict the effect of hydrophobicity, lubricant level and excipient brittleness on spironolactone dissolution, compact disintegration time and particle size. The ANOVA table for the cubic and quadratic models is shown in Table 6. However, the experimental data did not fit the surface models for drug dissolution (r^2= 0.5728). This is explained by the small variability of data for drug dissolution with lubricant levels and lubricant hydrophobicity. Even though all tablet batches contained a disintegrant, none of them was able to release 100% of spironolactone within 1 hour of study. This is explained by the hydrophobic nature of the drug which restricted dissolution.

Hydrophobicity and excipient brittleness, rather than the lubricant level were the significant factors that influenced compact disintegration and the resulting particle size of the powder mixtures (Figure 4). In fact, the fastest disintegrating compacts were produced with blends containing a high amount of stearic acid and lactose which were the less hydrophobic lubricant and the most brittle excipient, respectively. Once compacts disintegrated in distilled water, they disaggregated into the original primary compounding particles. As a result, compacts having a large amount of sorbitol exhibited the largest particle size which combined with its slow dissolving nature and high compact hardness delayed the disintegration time.

The cubic and quadr atic reduced models for disintegration time and particle size are expressed as follows:

$$\text{Disintegration time} = 471+104.6A-10.66B-211C+ 58.8AB\ 39.9AC +5.84B^2-192.8C^2-141.5AB^2$$

$$\text{Particle size} = 150.1-0.81A-54.33C-2.18AC+5.51C^2$$

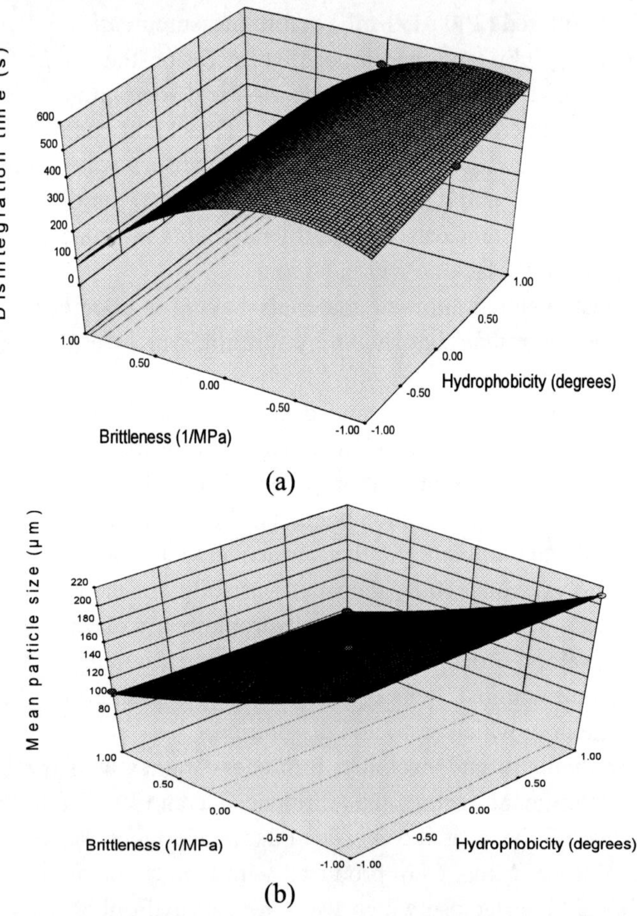

Figure 4. Surface plots of (A) Disintegration time (B) Mean particle size for spironolactone-excipient mixtures.

Table 6. ANOVA Table for the reduce cubic and quadratic models obtained for disintegration time and particle size, respectively

Response	Source	Sum of Squares	df	Mean Square	F-value	p-value
Compact disintegration time	Model	566037.2	8	70754.7	33.1	0.0002
	A-Hydrophobicity	43764.6	1	43764.6	20.5	0.004
	B-Concentration	909.5	1	909.5	0.43	0.539
	C-Brittleness	356168	1	356168	166.4	< 0.0001
	AB	13818	1	13818	6.5	0.044
	C^2	138119.2	1	138119.2	64.54	0.0002
	AB^2	40058.7	1	40058.7	18.72	0.005
	Residual	12841.2	6	2140.2		
	Lack of Fit	5918.2	4	1479.5	0.43	0.786
	Pure Error	6923.0	2	3461.5		
	Corrected Total	578878.4	14		r^2	0.9778
Mean particle size	Model	23747.4	4	5936.8	1018.7	< 0.0001
	A-Hydrophobicity	5.3	1	5.3	0.91	0.363
	C- Brittleness	23609.7	1	23609.7	4051.1	< 0.0001
	AC	18.9	1	18.9	3.23	0.102
	C^2	113.5	1	113.5	19.5	0.001
	Residual	58.3	10	5.8		
	Lack of Fit	33.8	8	4.2	0.34	0.888
	Pure Error	24.5	2	12.3		
	Corrected Total	23805.7	14		r^2	0.9976

The lubricant level had no effect on the particle size and was not significant for the compact disintegration time. The predicted combination of factors that rendered the shortest disintegration time and a particle size between 98 and 210 μm was achieved with stearic acid at a 1.5% level in mixtures with lactose monohydrate rendering a disintegration time and particle size of 13 s and 104 μm, respectively. These values are closed to the ones obtained from the validation run (23 s and 98 μm, respectively).

Conclusion

Lubricants eased the compaction process by reducing die-wall friction during tablet ejection, improving flowability, bulk and tap densities, compressibility and reducing the adhesion of the powder to metal surfaces. Since stearic acid and derivatives are hydrophobic, the formation of an external film on the host particles reduced surface wettability, decreased dissolution rates and prolonged disintegration times. Lubricants also weakened the bonding of the powder mixtures by creating an interface on the surface which reduced particle binding. These effects were aggravated by increasing the amount of lubricant and blending time, and by using mainly plastic-deforming and more regularly-shaped materials with a smooth surface such as sorbitol. Regarding tablet lubrication, the use of stearic acid led to the production of tablets with a higher compactibility and dissolution after 60 minutes as compared to magnesium stearate and sodium stearyl fumarate.

From lubricant applications to controlled released devices, the progress in stearic acid and derivatives are considered as challenging and ground-breaking. Future innovations in stearic acid-based lubricants could lead to the development of high-tech pharmaceutical excipients and devices which improve the production of pharmaceutical dosage forms.

Acknowledgments

The author thanks Mr. Edwin Marin of Laproff laboratories for proving us with the materials employed in this study. Mr. Yhors Ciro is also acknowledged for his technical assistance.

References

[1] Aoshima, H; Miyagisnima, A; Nozawa, Y; Sadzuka, Y; Takashi, Sonobe, T. Glycerin fatty acid esters as a new lubricant of tablets. *Int J Pharm*, 2005, 293, 25-34.

[2] Yüksel, N; Türkmen, B; Kurdoğlu, A; Başaran, B; Erkin, J; Baykara, Fabad T. Lubricant efficiency of magnesium stearate in direct compressible powder mixtures comprising Cellactose® 80 and pyridoxine hydrochloride. *J Pharm Sci*, 2007, 32, 173-183.

[3] Vitkova, M; Chalabala, M. The use of some hydrophobic substances in tablet technology. *Acta Pharm Hung*, 1968, 68, 336-44.

[4] Bastos, M; Friederich, R; Beck, R. Effects of filler-binders and lubricants on physicochemical properties of tablets obtained by direct compression: A 22 factorial design. *Lat Am J Pharm*, 2008, 27, 578-83.

[5] Koivisto, M; Jalonenb, H; Lehtoa, VP. Effect of temperature and humidity on vegetable grade magnesium stearate. *Powder Technol*, 2004, 147, 79-85.

[6] Roberts, M; Ford, JL; MacLeod, GS; Fell, JT; Smith, GW; Rowe, P; Dyas, AM. Effect of lubricant type and concentration on the punch tip adherence of model ibuprofen formulations. *J Pharm Pharmacol*, 2004, 56, 299-305.

[7] Wang, J; Wenb, H; Desai, D. Lubrication in tablet formulations. *Eur J Pharm Biopharm*, 2010, 75, 1-15.

[8] Okoye, P; Stephen, H. Lubrication of direct-compressible blends with magnesium stearate monohydrate and dihydrate: The influence of magnesium stearate (MgSt) on powder lubrication and finished solid-dose properties presents big challenges to drug manufacturers. *Pharm Technol*, 2007, 1, 1-12.

[9] Rojas, J. Excipient functionality enhancement: The cellulose II case. The application of the coprocessing technology on excipients (1st edition). Saarbrucken, Germany: Lambert Academic Press, 2012.

[10] Veiga, G; Ribeiro, TB; Pavie, LC; Souto, E; Pinho, S. Characterization and shelf life of β-carotene loaded solid lipid microparticles produced with stearic acid and sunflower oil. *Braz Arch Biol Technol*, 2013, 56, 663-671.

[11] Sharpe, SA; Celik, M; Newman, A; Brittain, G. Physical Characterization of the polymorphic variations of magnesium stearate and magnesium palmitate hydrates species. *Structural Chem*, 1997, 8, 73-84.

[12] Rojas, J; Yepes, M; Ciro, Y. Viscosity, Agglomeration Behavior and Tableting Characteristics of Cassava Starch Modulated by Wet Granulation: A Comparative Study. In F. P. Molinari (Eds), Cassava: Production, Nutritional Properties and Health Effects (1st. ed., 2014, 139-159). New York, NY: Nova Science Publishers.

[13] Rojas, J. Effect polymorphism on the powder and tableting properties of cellulose. In: T. Van de Ven, & L. Godbout (Eds.) Cellulose Medical, Pharmaceutical and Electronic applications (1st. ed., 2013, 27-46). Rijeka, Croatia: InTech.

[14] Rojas, J; Aristizabal, J; Henao, M; Sánchez, J. Screening of several excipients for direct compression: A new perspective based on functional properties. *Rev Cien Farm Bás Apl*, 2013, 34, 17-23.

In: Stearic Acid
Editors: Yunfeng Lin and Qiang Peng
ISBN: 978-1-63463-172-3

Chapter 5

DEUTERATED HYDROXYSTEARIC ACIDS AND THEIR APPLICATIONS

Gabriele Micheletti, Carla Boga*, Silvia Cino, Paolo Caruana and Paolo Zani*

Department of Industrial Chemistry 'Toso Montanari' Alma Mater Studiorum – Università di Bologna, Bologna, Italy

ABSTRACT

Deuterium labelled compounds possess a wide range of applications in pharmaceuticals, environmental, materials and chemical sciences. They are useful in the investigation of bio-active molecules, giving detailed insights into the mechanism of chemical reactions. The deuteration of fatty acids, including stearic acid and its derivatives, permits to obtain reference or standard compounds for many purposes, including studies on metabolic pathways and analytical determinations. This paper refers to hydroxystearic acids, a class of compounds of interest in many applied fields focusing particular attention on the most used deuterium-labelling methodologies and related problems. An outline on the main applications of deuterated hydroxystearic acids is also done, critically discussing the examples taken from the literature. Furthermore, new results are presented concerning a simple protocol for obtaining D_2 '*in situ*': using the seeds of *Dimorphoteca sinuata* L. as natural precursor, and a lab-scale

* Corresponding author: Email: Carla Boga, carla.boga@unibo.it.
* Corresponding author: Email: Paolo Zani, paolo.zani@unibo.it.

apparatus developed by us, we have verified the possibility to obtain deuterium-pluri-labelled 9-hydroxystearic acid.

INTRODUCTION

Fatty acids are naturally occurring compounds that usually consist of a carboxyl group bound to an hydrophobic, saturated or unsaturated, alkyl chain with more than 4 carbons: they play an essential role in biology. Fatty acids represent a stock of energy for the organisms, are constituent of membranes and oil seeds and are involved in many biological pathways. Furthermore, they are of growing interest in industrial and food chemistry. Besides unsubstituted fatty acids, there are several fatty acids bearing different functionalities along the chain that give them a peculiar role in biological pathways. Among these latter, hydroxy fatty acids (HSAs) and their derivatives represent a class of compounds of growing interest in many scientific areas. In industry, they are widely used as lubricants, [1-3] surfactants, [4, 5] plasticizers, [6, 7] additives in coatings and paintings, [8-10] components of detergents, [11] cosmetics, [12, 13] flavours, [14] and foods.[15, 16] In the biological field, they are ubiquitous and are mainly found in triacylglycerols, membrane phospholipids, waxes, cerebrosides and other lipids.[17-19] 2-(*R*)-hydroxy fatty acids occur in appreciable amounts in the sphingolipids of plants and 2- and 3-hydroxystearic acids are present in a wide range of microorganisms and are characteristic constituents of lipopolysaccharides and lipid A, located in the outer membrane of Gram-negative bacteria and essential to the endotoxin activity. [20] 10-Hydroxystearic acid (10-HSA) is the major component of adipocere, the waxy material produced by microbial decay of organic matter [21, 22] and is receiving growing importance as precursor of the γ-dodecalactone, in turn an important contributor to the aroma of several fruits and also to the sweet and fatty flavour of malt whisky. [23-25] HSAs have been found in membrane lipids subjected to oxidative stress. Recently, 9-HSA revealed to play an important role in cancerogenesis, [26] since hystone deacetylases (HDAC), [27] are its molecular target. HDAC is an enzymatic family that is considered a potential target for antitumor therapies. Recently, monolayers at the air-water interface of some monohydroxystearic acid (namely, 2-, 7-, 9-, and 12-HSA) have been studied and a specific correlation between the position of the secondary OH group and the crystalline symmetry of the monolayer has been demonstrated, [28] thus confirming the important role played by this family of compounds also in material chemistry. Furthermore, 17-HSA is an important constituent of beeswax. [29] Valuable natural sources of optically pure saturated and/or unsaturated HSAs are the seed oils of higher plants [30] such as Proteaceae, Compositae, Euphorbiaceae, Cruciferae, Labiateae, Oleacaceae, etc. The best known of these is ricinoleic acid (D-(-)12-hydroxy-octadec-*cis*-9-enoic acid), which comprises up to 90% of castor oil (from *Ricinus communis*) and is also the major component of the lipids of the fungus ergot (*Claviceps purpurea*). A convenient route to (*R*)-9-HSA starts from 9-hydroxy-octadeca-*trans*-10, *trans*-12-dienoic acid ((*S*)-dimorphecolic acid) present in large amount in the seed oil of genus *Dimorphotheca*. [30] Coriolic acid (13-Hydroxy-octadeca-*cis*-9, *trans*-11-dienoic acid), available from *Coriaria nepalensis* seed oil, [30] is a natural source of the unusual D-13-HSA.

Deuterated hydroxystearic acids (HSAs) are the subject of this review, and their synthesis is closely related to their uses and applications in different research fields.

For this reason, and for sake of clarity, the syntheses and uses of deuterated HSA (including their methyl esters) will be discussed in separate sub-headings, depending on the field in which they have found application.

1. General Synthetic Protocols to Obtain Deuterated Hydroxystearic Acids

In the following sub-headings the most common synthetic strategies to obtain HSAs specifically deuterated in different position of the alkyl chain are summarized. Other particular cases will be reported in the sections related to the use of specific HSAs.

1.1. Synthesis of 2, 2-d$_2$-HSAs

The introduction of two deuterons on the α-carbon of hydroxystearic acids can be easily accomplished by heating the methyl hydroxyester with NaOD in D_2O at high temperature and for long time (often in autoclave at 200 °C for 3 days), as depicted in Figure 1. For example, in this manner, after neutralization and esterification of the obtained product with methanolic hydrogen chloride, Tulloch [31] had synthesized a series of compounds, in particular methyl 6-, 7-, 8-, 9-, 10-, 11-, 12-, and 13-HSA-2,2-d_2.

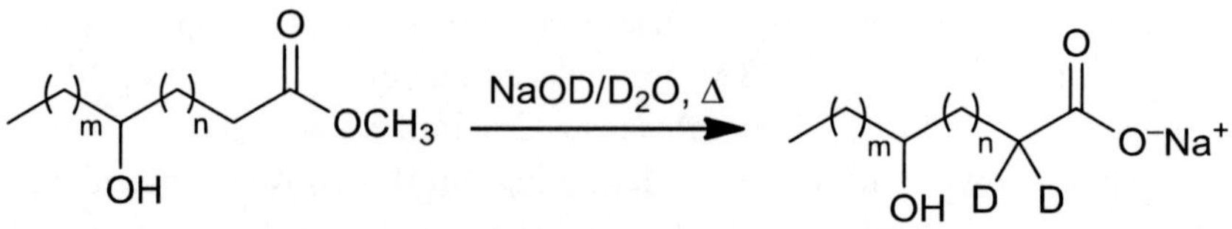

Figure 1. General synthesis of α,α-d_2-HSAs.

1.2. Deuteration on Carbon Atom Bearing the Hydroxyl Group

The most simple and used mehod to obtain HSA carrying a deuterium on the carbon atom bound to the hydroxy group, is to start from the corresponding oxo-ester and to reduce it with $NaBD_4$ at room temperature (Figure 2) [27,31] or with $NaBD_3CN$. [32]

The ketoester is, in turn, usually synthesized beginning from the monoester of a dicarboxylic acid that is transformed into the acyl chloride by reaction with thionyl chloride (or oxalyl chloride) in anhydrous ethyl ether. Then the synthesis proceeds reacting the mono ester mono acyl chloride of the suitable dicarboxylic acid with an organometallic reagent (Figure 3). In the past, cadmium reagents were employed but, due to their toxicity, now they have been replaced by the more safe Grignard reagents (Figure 3). The Grignard reagents were used in anhydrous THF at -78 °C for a few minutes [33] or also in the presence of CuI a room temperature for 2 h.[34]

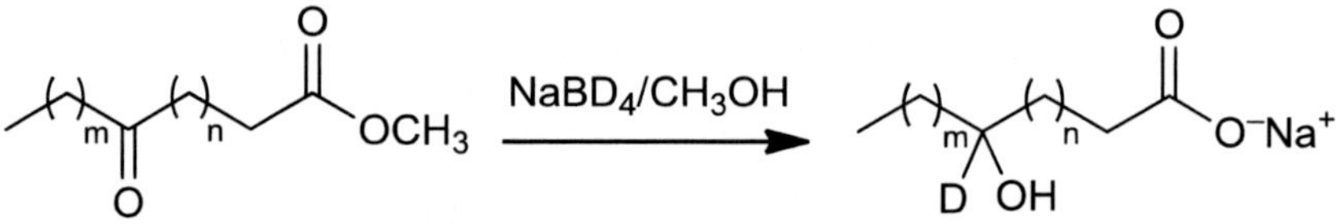

Figure 2. Synthesis of HSAs deuterated on the carbon atom bound to the hydroxyl group.

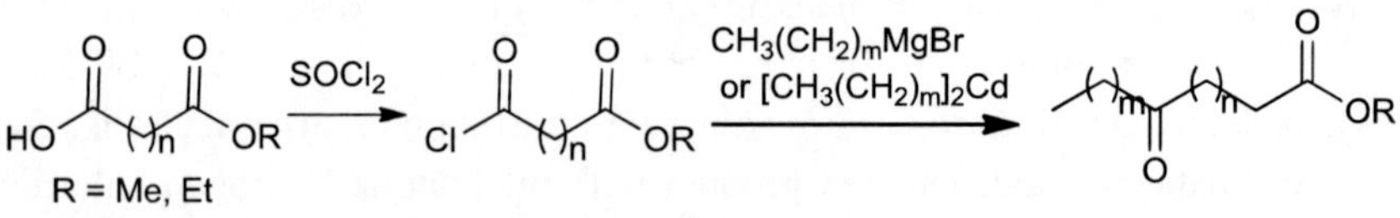

Figure 3. Procedure to obtain keto esters precursors of specifically HSAs functionalized at a specific position along the chain.

1.3. Deuteration on Carbon Atoms Adjacent to the Hydroxylated Position

The deuteration in α,α' positions with respect to the carbon atom bearing the hydroxyl group is usually performed by hydrogen-deuterium exchange on the keto-derivatives, as shown in Figure 4. [35]

This approach, shown in Figure 4, [36] has been followed to prepare numerous deuterated stearic acids. Actually, the α,α'-deuterated hydroxystearate is often transformed into the *p*-toluenesulfonyl derivative that, in turn, was treated with lithium aluminum hydride to obtain the deuterated octadecanol. Oxidation of this latter give deuterated stearic acid. If lithium aluminum deuteride is used instead of $LiAlH_4$, nucleophilic substitution of the tosylate with deuterium can be accomplished.

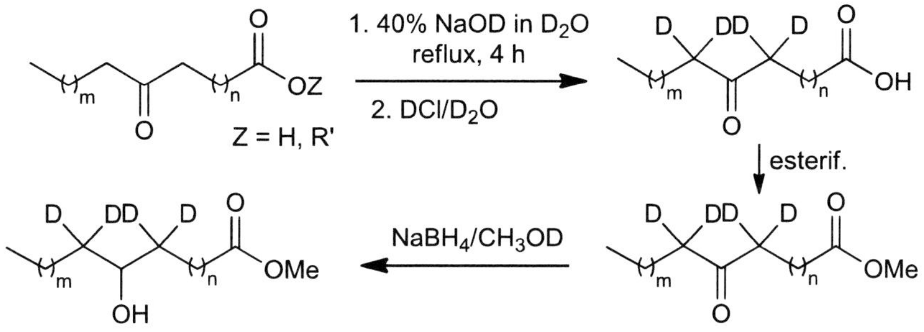

1. 40% NaOD in D_2O reflux, 4 h
2. DCl/D_2O

Z = H, R'

esterif.

$NaBH_4/CH_3OD$

Figure 4. Synthesis of HSAs deuterated in positions adjacent to hydroxy group.

1.4. Catalytic Reduction with Deuterium of Unsaturated HSAs

The introduction of deuterium atom by deuteration of double or triple bonds of unsaturated hydroxy fatty acids has been reported [29, 36] without particular comments on the deuterium incorporation.

However, despite the fact that the catalytic hydrogenation of double or triple bonds is a well-known and widely used reaction, several authors reported that when deuteration of unsaturated fatty acids or seed oils is carried out with deuterium instead of hydrogen, deuterium scrambling along the aliphatic chain and incorporation of a number of deuterium atoms higher than those required by the saturation of the double bonds can sometimes occur. [37] This extensive incorporation of deuterium atoms was explained in terms of double bond shift and isomerization consistent with an extension of the Horiuti-Polanyi mechanism. [38]

This phenomenon can occur in deuterations carried out in the presence of Palladium, Platinum, Nickel, Copper-chromite, [39] and also with Wilkinson's catalyst, if not freshly prepared. [40]

Since some authors did not report information on the above cited side reactions, we carried out some catalytic deuterations of methyl dimorphecolate (methyl 9-hydroxy-octadeca-*trans*-10, *trans*-12-dienoate), an optically pure precursor of (9*R*)-9-HSA. We used an 'in situ' deuterium production method, and we report the new results obtained in section 4 of this Chapter.

2. Applications of Deuterated Hydroxystearic Acids

The main applications of deuterated hydroxystearic acids are in the field of spectroscopy (mainly mass and ^{13}C NMR) and in the biological one (as intermediates to prepare labelled lipids or as key substrates to elucidate or gain information on metabolic pathways).

2.1. Nuclear Magnetic Resonance Investigations

The introduction of one or more deuterium atoms at specific carbons are extremely useful in NMR investigations, especially in the assignement of ^{13}C NMR resonances in related spectra. In this regard, a pioneeristic and basilar work has been made by Tulloch in 1970's. In a first paper, [41] he studied a series of deuterium labelled C18 aliphatic esters observing, besides that on the α-carbon, an upfield isotope shifts of signals for carbons situated in β and γ to the deuterium, and an increase of the line-width attributed to long range ^{13}C-D coupling. In that paper, methyl 17-HSA-17-*d* and methyl 17-HSA-16,16,18,18,18-d_5 were synthesized and used as comparison compounds. The first was obtained by reduction of methyl 17-oxooctadecanoate [36] as described in Figure 2, whereas the second was produced [36] through the procedure depicted in Figure 3, followed by esterification. These compounds have been also used to obtain the corresponding stearic esters after substitution of the hydroxyl with hydrogen. From this study it emerged that the ^{13}C shift of the β-carbon is smaller in case of hydroxylated compound (methyl 17-HSA-

17-*d*) with respect to that of methyl 17,17-d_2 octadecanoate. In methyl 17-HSA-16,16,18,18,18-d_5 the upfield shift was about half than that observed in the parent methyl 16,16,18,18,18-d_5 octadecanoate.

A further NMR study [42] was focused on the effect of the hydroxyl group on the ^{13}C NMR chemical shift of a series of methyl HSAs. For this purpose, sixteen specifically deuterated hydroxyesters and their acetates were employed and the analysis of the deuterium isotope effect on the spectra permitted to unambiguously assign the signals. Two series of labelled derivatives have been used in this study: 2,2-dideuterated methyl 6-, 7-, 8-, 9-, 10-, 11-, 12-, and 13-HSA, prepared as depicted in Figure 1, and then esterified with methanolic hydrogen chloride, and/or γ- or γ'-dideuterated (respect to the hydroxyl position) HSA methyl esters. In particular, they were: methyl 7-HSA-10,10-d_2, methyl 8-HSA-5,5-d_2, methyl 8-HSA-11,11-d_2, methyl 11-HSA-8,8-d_2, methyl 11-HSA-14,14-d_2, methyl 12-HSA-9,9-d_2, and methyl 13-HSA-16,16-d_2. They were prepared by reduction with $NaBH_4$ of the corresponding gamma- or gamma'-deuterated oxoester precursors. In turn, these latter were synthesized through different synthetic approaches. Oxo-acids with the dideuterated carbon situated on the side of the terminal CH_3 chain were obtained through 2 routes, as shown in Figure 5.

The synthesis begins from a 2,2-dideuteroacid that, after reduction into the corresponding alcohol and transformation into mesilate, undergoes malonic synthesis. In the *Route a,* the substituted acetic acid obtained after decarboxylation was transformed into acyl chloride and subjected to enamine synthesis according to Hunig's methodology giving, after work-up, 2-ketocycloalkanones which, with aqueous alkali, lead to homologated ketoacids. [43, 44] Thus, from hexanoyl-4,4-d_2 chloride and morpholinocyclododecene (see Figure 5, x = 7), 13-oxo-octadecanoic-16,16-d_2 was formed, whereas 7-oxo-octadecanoic-10,10-d_2 was obtained from dodecanoyl-4,4-d_2 chloride and morpholinocyclohexene (see Figure 5, x =1).

From 2,2-dideuteroalkylmalonic acids and half ester acid chlorides *(Route b)* 8-oxo-octadecanoic-11,11-d_2 and 11-oxo-octadecanoic-14,14-d_2 acid were obtained.[45]

Oxo-octadecanoic acids dideuterated in γ-position on the carboxyl side were prepared from 4,4-dideuterated ω-unsaturated acids as shown in Figure 6, by acylation of substituted malonic acid, functional group manipulation; the last step was the oxidation of the terminal methylene unit to carboxylic acid.

A recent application [34] of deuterated hydroxyacids in NMR spectroscopy has been the study of lipid rafts in membranes. It is thought that sphingomielin and cholesterol are basilar for forming lipid rafts.

RCD_2CH_2OH

↓ MsCl

RCD_2CH_2OMs malonic synthesis

route a → $RCD_2(CH_2)_2CO(CH_2)_nCOOH$ {R = Et, n = 11 (13-oxo-16,16-d_2); R = n-octyl, n = 5 (7-oxo-10,10-d_2)}

route b → $RCD_2(CH_2)_2CO(CH_2)_nCOOH$ {R = n-heptyl, n = 6 (8-oxo-11,11-d_2); R = n-Bu, n = 9 (11-oxo-14,14-d_2)}

Route a

$RCD_2CH_2CH_2COOH$

↓

$RCD_2CH_2CH_2COCl$ + (enamine, morpholine, $)_x$) → $RCD_2CH_2CH_2$ (2-acylcycloalkanone, $)_x$) $\xrightarrow{OH^-}$ $RCD_2CH_2CH_2CO(CH_2)_3(CH_2)_xCH_2COOH$

Route b

$RCD_2CH_2CH(COOH)_2$ + $ClCO(CH_2)_nCOOCH_3$ → $RCD_2(CH_2)_2CO(CH_2)_nCOOH$

Figure 5. Synthesis of oxo-octadecanoic acid derivatives gamma-deuterated on the side of the terminal CH_3 chain.

$CH_2{=}CH(CH_2)_nCD_2CH_2CH_2COCl$

+

$CH_3(CH_2)_{m-1}CH(COOH)_2$

→ $CH_2{=}CH(CH_2)_nCD_2(CH_2)_2CO(CH_2)_mCH_3$

↓ [O]

$HOOC(CH_2)_{n+1}CD_2(CH_2)_2CO(CH_2)_mCH_3$

m = 6, n = 6 (11-oxo-8,8-d_2)
m = 3, n = 9 (8-oxo-5,5-d_2)
m = 7, n =5 (12-oxo-9,9-d_2)

Figure 6. Preparation of oxo-octadecanoic acids dideuterated in gamma-position on the carboxyl side.

NMR technique, especially solid state ^{2}H NMR spectrometry, has been frequently used in investigations of the order and mobility of acyl chains within the lipid bilayers. The ^{2}H NMR study of the quadrupole splitting of deuterons located along the lipid chain permitted to gain information on molecular motions of lipids. For this aim, selectively deuterated lipids have been prepared. *N*-stearoyl sphingomyelin (Figure 7) was deuterium-labelled on different positions of both the sphingosine and the stearoyl chains to capture its motion in membranes.

Methyl esters of 6-, 8-, 10-, 12-, 14-, and 16-HSA deuterated on the carbon atom bearing the hydroxyl group have been prepared and used to functionalize selectively with deuterium atoms some sphingomielins. They were prepared by reduction with $NaBD_4$ of the corresponding ketoester, in turn obtained as previously reported in Figure 3.

A further study, [46] aimed to elucidate the conformation of sphingomielins in membranes utilized bicelles as a model for the lipid bilayer. ^{31}P NMR, ^{1}H NMR, ^{2}H NMR, and dynamic light scattering experiments gave information on the orientation and isotropy of bicelles. In this case, 10-HSA-10-*d* was used as intermediate to obtain sphingomielins containing the stearoyl moiety dideuterated in positon 10. For this purpose, methyl 10-hydroxystearate-10-*d*, prepared as described in Figure 2, was firstly transformed in tosylate then treated with sodium borodeuteride to give the saturated methyl stearate-10,10-d_2 that, in turn, was hydrolyzed, transformed into the more reactive p-nitrophenyl ester, and bound to amino group of sphingosine-1-phosphorylcholine through acylic nucleophilic substitution.

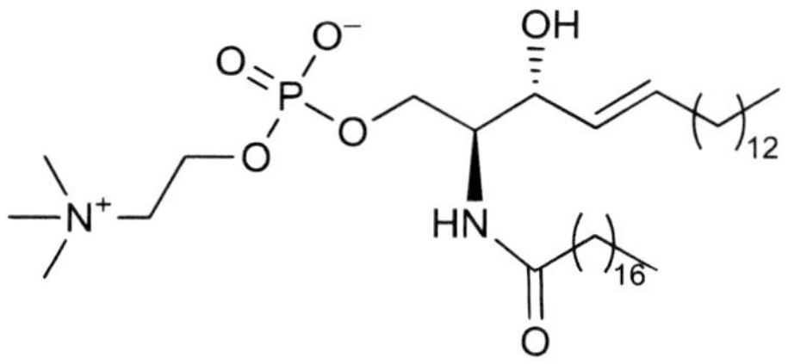

Figure 7. *N*-stearoyl sphingomyelin.

2.2. Mass Spectrometry Determinations

The quantitative determination by mass spectrometry of 2- and 3-hydroxy fatty acids in complex food matrices and as chemical markers for the detection of endotoxins derived from Gram-negative bacteria requires their transformation into the corresponding methyl esters and, often, the additional derivatization of hydroxyl groups in trimethylsilyl ethers or other derivatives. In a research work on the content of 2- and 3-HSAs in bovine milk, deuterated hydroxy fatty acids have been particularly useful as internal standards for mass analyses.[20] For this study, the methyl ester of 2-HSA-9,10-d_2 was prepared from methyl 2-hydroxy-9-cis-octadecenoate by deuteration in the presence of Wilkinson's catalyst. The hydroxy group was transformed into the pentafluorobenzoyl ester and the product was used as standard in gas chromatography with electron-capture negative ion mass spectrometry.

Rontani and Aubert [47] studied the EI mass spectral fragmentation of hydroxy carboxylic acids and trimethylsilyl derivatives of oxocarboxylic acids. Some fragmentations appeared to be specific to characterize and distinguish HSAs trimethylsilylated on both the hydroxy and the carboxy functionality, from the corresponding trimethylsilyl keto acids. Some ions have been rationalized hypothesizing hydrogen and trimethylsilyl transfer and this was corroborated by EI-MS experiments carried out using the trimethylsilyl derivatives of 10-HSA-10-*d*, 11-HSA-11-*d*, 12-HSA-12-*d*, and 10-ketostearic acid-9,9,11,11-d_4. The first three were obtained according to the method depicted in Figure 2, whereas the latter was obtained by hydrogen-deuterium exchange (Figure 4). Also a mixture of 11-HSA-12-*d* and 12-HSA-11-*d* was used. Its synthesis was accomplished by epoxidation of vaccenic acid (*vide infra*) and reduction of the epoxide with $NaBD_4$.

A more recent application of 9-HSA-9-*d* was in liquid chromatography with electrospray ionization-mass detector (LC-MS) analyses. [27] In particular, 9-HSA, an endogeneous lipid peroxidation by-product, showed to play a key role in controlling the cell division and is present in minor amount in tumor cells with respect to normal cells. When 9-HSA is administered exogenously to HT29 cells, a cell line of human colon tumor, it induces the expression of a cell cycle kinase inhibitor. The use of 9-HSA-9-*d* for tracing calibration graphs resulted to be essential during a study aimed at depeen this behavior, that was correlated to the inhibition of hystone deacethylase. Also in this case, the labelling was carried out by reduction of the 9-keto precursor through the approach shown in Figure 2.

2.3. Biochemical Investigations

2.3.1. Biosynthesis of Mandibular Acids in Honey Bees [29]

Mandibular glands of workers and queens of honey bees (*Apis mellifera* L.) contain predominantly 10-carbon functionalized fatty acids and their function is closely related to the respective tasks in the colony. Those produced by queens and workers differ for the position of the functionalization: the first bear a functional group at the ω-1 position, whereas the latter are functionalized in ω-position. Their biosynthesis starts from stearic acid and proceeds through hydroxylation at the ω-1-or ω-position, chain cleavage, and oxidation of the hydroxyl group to give diacids and 9-ketodecenoic acid. In this context, the synthesis of deuterium-labelled hydroxyoctadecanoic acids, such as 18-HSA-18-*d* and 17-HSA-17,18,18-d_3 and their incorporation into mandibular glands has permitted to deepen the mechanism of interconversion between hydroxyacids.

18-HSA-18-*d* was synthesized from the monoacetate of 1,18-octadecandiol: through oxidation followed by saponification, the resulting 18-oxoacid was reduced with $NaBD_4$ producing the ω-deuterated HSA (Figure 8).

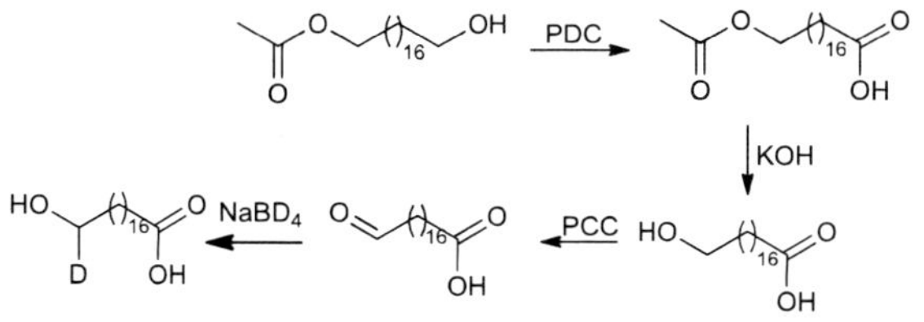

Figure 8. Synthesis of 18-HSA-18-*d*.

More laborious was the synthesis of 17-HSA-17,18,18-d_3, as shown in Figure 9 [29]:

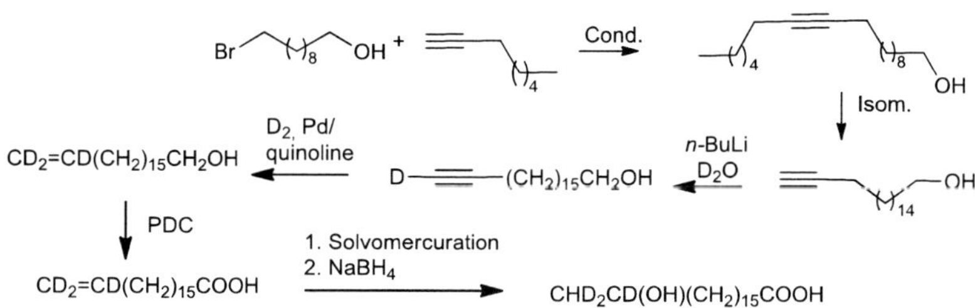

Figure 9. Synthetic pathway to 17-HSA-17,18,18-d_3.

From this study it has been observed that worker bees glands readily shorten 18-HSA-18-*d* into 12-HSA-12-*d* whereas in queens 17-HSA-17,18,18-d_3 was transformed into the deuterium-labelled (ω–1)-hydroxylated 12-HSA. In addition, worker's glands elongate 16-hydroxypalmitic-15,15-d_2 acid to 18-HSA with deuterium incorporated.

2.3.2. Yeast's Metabolites

The stereospecific biochemical hydroxylation of C18 unsaturated fatty acids in a species of yeast belonging to the genus *Torulopsis* occurs when it was grown on a medium containing glucose and long chain fatty acids. [36]

After hydroxylation, the hydroxy fatty acids, directly derived from the administered fatty acids, are converted into glycosides. It has been found that L-17-hydroxyacid is produced by the yeast when stearic or oleic acid were supplemented. The investigation on the mechanism of the hydroxylation required the administration to yeast cultures of labelled substrates, including deuterium-labelled fatty acids. [36]. In particular, methyl octadecanoate-17,17-d_2 and ethyl stearate-16,16,18,18,18-d_5 were required. Their synthesis started from methyl 17-oxooctadecanoate and involved deuterated 17-HSAs as intermediates. Methyl 17-HSA-17-*d* was obtained as shown in Figure 2, then it was subjected to tosylation, reduction with lithium aluminum deuteride, oxidation of the primary alcoholic functionality and esterification. Ethyl stearate-16,16,18,18,18-d_5 was prepared through the following steps: α,α‘-deuteration of methyl 17-oxooctadecanoate with the protocol shown in Figure 4, conversion of the obtained 17-oxostearic acid-16,16,18,18,18-d_5 into methyl ester, reduction of the keto group to hydroxy group with $NaBH_4$, tosylation of the hydroxyl, reduction to methylene group with $LiAlH_4$, oxidation with chromiun trioxide of the primary alcohol formed by reduction of the carboxyl, and final esterification by usual methods. From the biological study it emerged that no deuterium atoms at C-16 and C-18 position were removed and that L-17-hydroxystearic acid is produced by displacement of an hydrogen atom with retention of configuration.

In another paper by Albrecht et al., [48] the biosynthetic pathway involved in the production of aliphatic lactones was studied. In particular, the study was focused on the mechanism of the biotransformation of (*S*)- and (*R,S*)-13-hydroxy-(*Z,E*)-9,11-octadecadienoic acid into the optically pure (*R*)-δ-decalactone by the yeast *Sporobolomyces odorus*. For this purpose, deuterium-containing derivatives of coriolic acid needed to be synthesized. Figure 10 shows the synthetic approach chosen by the authors to produce 13-HSA-13-*d* and 13-HSA-9,10,11,12-d_4 starting from linoleic acid.

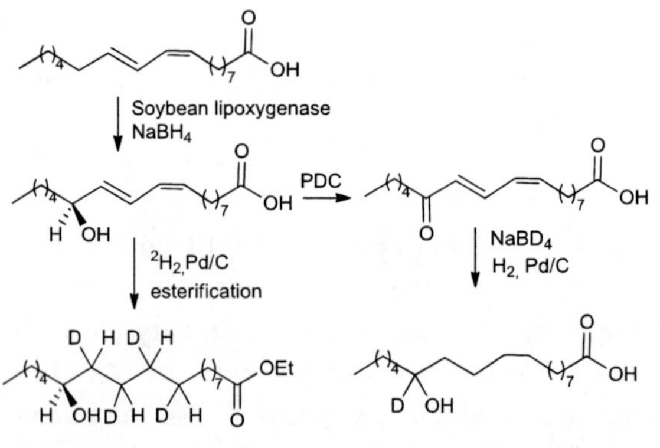

Figure 10. Synthesis of 13-HSA-13-*d* and 13-HSA-9,10,11,12-d_4 starting from linoleic acid.

The absence of deuterium in the δ-decalactone obtained starting from 13-HSA-13-*d*, together with the results obtained with the tetradeuterated analogue and its 13-ketoderivative support a mechanism involving the oxidation of the hydroxy group followed by an enantioselective reduction of the keto group thus obtained.

3. Deuterated Dihydroxystearic Acids

12,13-Dihydroxyoctadecanoate-9,10-d_2 has been chosen as starting material for the synthesis of methyl 12-oxo-dodecanoate-9,10-d_2, an intermediate to obtain labelled octadecanoic and octadecenoic acids. Wittig reaction between 12,13-dihydroxyoctadecanoate-9,10-d_2 and *n*-hexyl-3,3,4,4-d_4-triphenylphosphonium bromide gave *cis*- and *trans*- hexadeuterated methyl 12-octadecenoates, useful for metabolism studies in humans. [49] The starting deuterated di-hydroxystearic acid was prepared from methyl *threo*-12,13-dihydroxy-*cis*-9-octadecenoate derived from *Vernonia anthelmintica* seed oil. This oil contains a large amount of vernolic acid, namely 12,13-epoxy-*cis*-9-octadecenoic acid, from which, after acetolysis, saponification, and esterification, methyl 12,13-dihydroxyoleate was obtained. Deuteration of this latter in the presence of Wilkinson's catalyst, followed by cleavage with red lead in glacial acetic acid lead to 12-oxo-dodecanoate-9,10-d_2.

The preparation of methyl *trans*-12-octadecenoate-9,10-d_2 was simpler. In this case methyl *threo*-12,13-dihydroxy-*cis*-9-octadecenoate was transformed into 12,13-di-*O*-(ethoxymethylene)-9-octadecenoate, then deuterated in the presence of Wilkinson's catalyst and finally converted into the 12-ene-derivative by heating at 210 °C. The deuterated target compound thus obtained was then transformed into glycerides for metabolic studies.

Further studies on this topic required the preparation of the four geometrically isomers of methyl 12,15-octadecadienoates-9,10-d_2. [50, 51] They were prepared by Wittig reaction between *cis*- or *trans*-3-hexenyl-triphenylphosphonium bromide and methyl 12-oxododecanoate-9,10-d_2 prepared from vernolic acid as above described.

Starting from oleic acid, [52] 9,10-dihydroxy-8,8,11,11-d_4 was obtained after oxidation to 9,10-dioxo derivative followed by hydrogen-deuterium exchange as shown in Figure 4 and reduction of he keto groups to hydroxy groups. The product obtained was converted to methyl oleate-8,8,11,11-d_4; in turn, the latter was used, together with others deuterium-labelled compounds, such as oleic acid-9,10-d_2 and *L*-methionine-methyl-d_3, to elucidate the mechanistic biochemical pathway involved in the cyclopropanation from *cis*-vaccenic acid and oleic acid to lactobacillic acid and dihydrosterculic acid, respectively (Figure 11); this cyclopropanation reaction occurs in *Lactobacillus plantarum* cultures.

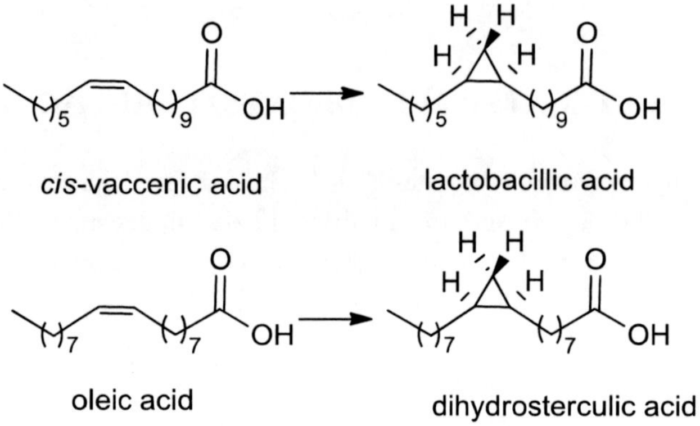

cis-vaccenic acid → lactobacillic acid

oleic acid → dihydrosterculic acid

Figure 11. Trasformation of unsaturated acids to cyclopropyl derivatives.

4. Our New Results: A Simple Lab-Scale Synthesis of Pluri-Deuterated 9-Hydroxystearic Acid

The availability of deuterated compounds containing no exchangeable deuterium atom is of great importance in biological studies. Actually, these compounds show the same interactions with cells of carbonium-hydrogen analogs, but their different molecular weight permit to distinguish them from the endogeneous products (if contained) by mass spectrometry.

In the past, we have synthesized some deuterium-labelled compounds as internal standards for mass spectrometric determinations of the endogenous content in normal and in tumour cells of short- and long-chain lipid peroxidation products. [27, 53, 54] Among them, 9-hydroxystearic acid (9-HSA) containing a C-D bond in position 9 was synthesized by reduction of the 9-keto precursor with $NaBD_4$. 9-HSA is biologically relevant as natural negative regulator of tumour cell proliferation. [27] It was found in human and murine epithelial cell lines, and acts as a competitive inhibitor on at least one human histone deacethylase (HDAC) isoform, HDAC1. [26] Prediction by molecular docking [27] indicates a favourable formation energy of the HDAC1-9-HSA complex and an higher stability of bonding for the enantiomer

(*R*)-9-HSA with respect to that with opposite configuration: this was confirmed by experiments. [55] This implied the synthesis of both enantiomers of 9-HSA in enantiopure form. Due to the quasi-symmetry of the moieties around the asymmetric carbon atom, the availability of enantiopure forms through classical enantioselective methods is unpracticable, as well as the kinetic resolution approach by means of enzymes. [56]

Enantiomerically pure (9*R*)-HSA can be obtained from *Dimorphoteca sinuata* L. seed oil, a source of 10(*E*),12(*E*)-(*S*)-9-hydroxyoctadecadienoic acid (dimorphecolic acid). [57] After transmethylation of the seed oil, followed by hydrogenation and hydrolysis, (9*R*)-HSA can be obtained and the enantiomer with opposite configuration can be synthesized through stereospecific inversion of the chiral center. [55]

The availability of (*R*) and (*S*) 9-HSA starting from natural oil containing dimorphecolic acid suggested us a method to achieve the polydeuterated analogue of 9-HSA. In fact dimorphecolic acid contains two coniugated double bonds and it might be possible to reduce them with molecular deuterium to synthesise the tetradeuterated product.

As above cited, monodeuterated compound (containing a C-D bond at position 9) of racemic 9-HSA was already obtained by reduction of the corresponding ketone with $NaBD_4$, [27] but it is important to underline that the higher is the number of deuterium atoms in the molecule, the easier is the recognition of the product in biological samples. For this reason, we planned to prepare (9*R*)-HSA labelled with up to four deuterium atoms. However, the importance to have 9-HSA in enantiopure form ruled out the possibility to label it in position 9, as previously made.

It is known that the catalytic reduction with deuterium of some mono- and polyunsaturated fatty acids (or their methyl esters) suffers redistribution (scrambling) and polydeuteration phenomena with production of isotopologues (see part 1.4). Nevertheless, in our opinion, this approach deserves to be tried owing to the fact that other methods, such as reduction with deuterium by deuteriodiimide [58] resulted no less laborious and dangerous.

In addition, since there are drawbacks to utilize deuterium gas on a laboratory scale, due to difficulties in the availability of small amount of deuterium and the handling of the gas itself, we thought that an optimal approach could be the '*in situ*' production of deuterium. The reduction step might be an easy way to obtain tetra-deuterated (9*R*)-9-HSA, even if we did not know, *a priori*, what might happen (scrambling, production of isotopologues) during the reduction process.

Herein we report the results obtained during the investigation of different 'lab-scale' procedures for the preparation, under mild conditions, of pluri-deuterated (9*R*)-9-HAS from *Dimorphoteca sinuata* L. seeds (Figure 12).

Since it is known that sodium borohydride slowly reacts with water to develop hydrogen, and it is commercially available also in deuteride form, we first planned to try the reduction of methyl (*S*)-dimorphecolate by producing molecular deuterium '*in situ*' from $NaBD_4$ and D_2O ($NaBD_4 + 2\ D_2O \rightarrow NaBO_2 + 4D_2$). Furthermore, sodium borodeuteride is a solid compound, safely stored and available in a powder form that allow quick production of small predetermined amounts of deuterium (D_2). However, it should be kept in mind that sodium borohydride itself and its reaction with water are not completely harmless and the reaction must be carried out with the appropriate precautions.

A first test was carried out in deuterated methanol as solvent and in the presence of the catalyst employing the raw material obtained after the transmethylation step shown in Figure 12. To this mixture, D_2O was slowly added by mean of a dropping funnel. Both the Adam's catalyst (PtO_2) and 10% Pd over charcoal resulted to be efficient, but the first is preferable due to the easier removal of the catalyst and work-up of the reaction. After this crucial step, the free acid was obtained by base-catalyzed hydrolysis followed by acidification of the carboxylate, as shown in Figure 12.

After that, we tried to produce *in situ* deuterium from the reaction between lithium and D_2O. In this case we used the Aldrich® diazomethane-generator (Aldrich code: Z411736 Aldrich) that resulted to be suitable for our purpose. In this apparatus, the mixture containing methyl dimorphecolate in methanol was put in the external tube, while in the internal tube, which contains lithium, we dropped D_2O with a siringe to produce deuterium. In this manner we were able to obtain, after purification by TLC, deuterated methyl hydroxystearate. However, this method suffers the limitation that no more than 20-30 mg of product can be obtained for a single batch.

In order to improve the protocol and to deuterate higher amounts of unsaturated material, we designed the deuterium production apparatus shown in Figure 13 that was then built in a Pyrex glass suitable to resist moderate pressure (as the glass used for bottles in Parr hydrogenator).

The apparatus is composed of two main parts: (1) an external vessel in which the hydrogen source (for either H_2 or D_2) is introduced, equipped with a D_2O (or H_2O) injection system; (2) an internal vessel containing the solvent, the hydrogenation catalyst, and the substrate to be deuterated (hydrogenated).

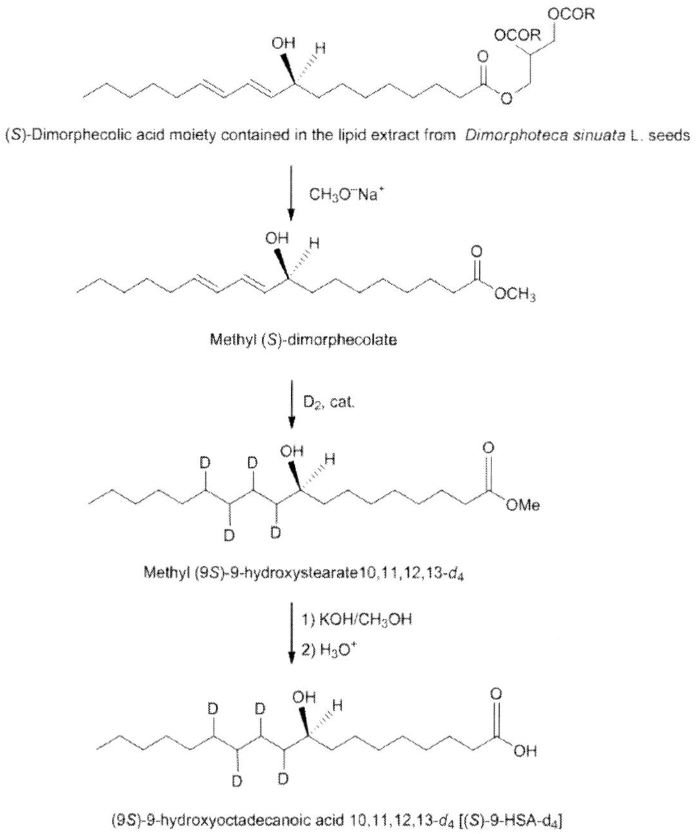

OCOR
OCOR
OH
H
O
O
(S)-Dimorphecolic acid moiety contained in the lipid extract from Dimorphoteca sinuata L. seeds
CH3O⁻Na⁺
OH
H
O
OCH3
Methyl (S)-dimorphecolate
D2, cat.
D
D
OH
H
O
OMe
D
D
Methyl (9S)-9-hydroxystearate10,11,12,13-d4
1) KOH/CH3OH
2) H3O⁺
D
D
OH
H
O
OH
D
D
(9S)-9-hydroxyoctadecanoic acid 10,11,12,13-d4 [(S)-9-HSA-d4]

Figure 12. Synthesis of (9*R*)-9-HSA-10,11,12,13-d_4.

Syringe to drop D2O (or H2O)
D2 (or H2) source and magnetic stirrer
Methanolic solution of compounds to be deuterated, Adam's catalyst, magnetic stirrer
Screw cap

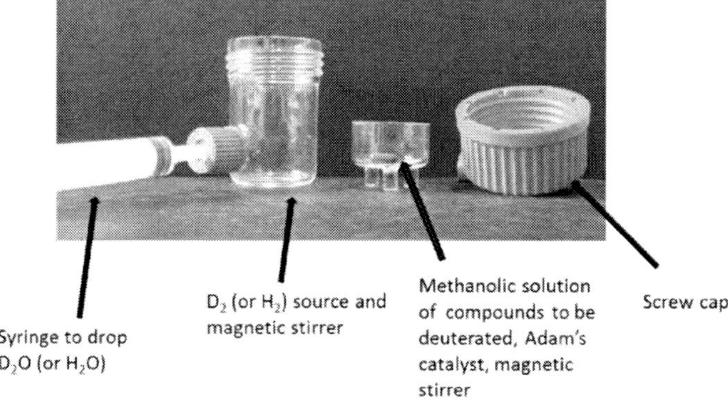

Figure 13. Apparatus designed by us for '*in situ*' production of deuterium (or hydrogen).

This small vessel is put inside the first one and is raised by bottom supports in order to prevent contact with the solution where D_2 (or H_2) is generated. Once closed with the screw cap, D_2O (or H_2O) was injected at the bottom of the first vessel. In this manner, deuterium (or hydrogen) thus produced reaches the reactant and the catalyst.

With this system we were able to deuterate, in a single batch, at room temperature and at the moderate pressure produced by gas evolution, about 300 mg of the crude mixture obtained after transmethylation of the lipid extract from *Dimorphoteca sinuata* L. seeds. The NMR analyses of the methyl ester and of the acid, especially ^{13}C NMR, showed the presence of deuterium atoms bound to the previously unsaturated carbons, as those showed typical ^{13}C-^{2}H coupling constants. [59] However, the spectra showed complex patterns, so we cannot exclude the presence of isotopologues. Mass spectra (ESI$^+$ mode) showed an highest peak corresponding to the tetradeuterated compound, even if an isotopic pattern typical of plurideuterated compounds is also visible.

Experimental Procedures

Oil extraction from *Dimorphoteca sinuata* L. seeds and transmethylation process were carried out as previously reported. [55]

Procedure A: under argon atmosphere, 50 mg of the transmethylated extract were dissolved in 1 mL of CD_3OD (sometimes 1 mL of ethyl acetate was added) then D_2O (1 mL) and 40 mg of $NaBD_4$ were added. To this mixture, 20 mg of Adam's catalyst were carefully added (caution!). When no more hydrogen is evolving, the reaction mixture was filtered, the solvent was removed and the crude reaction mixture was analyzed by ^{1}H NMR. The absence of vinylic proton signals revealed the reaction was completed.

After purification by flash chromatography (eluent: light petroleum/diethyl ether 7:3), 20 mg of pure deuterated methyl 9-hydroxy-stearate were recovered: ^{1}H NMR (600 MHz, $CDCl_3$): 3.66 (s, 3H, OCH_3), 3.60-3.55 (m, 1H, CHOH), 2.30 (t, 2H, J = 7.9 Hz, CH_2COO), 1.65–1.50 (m, 3H, CH_2CH_2COO and OH), 1.50–1.15 (m, 22 H), 0.88 (t, 3H, J = 6.7 Hz, CH_3). ^{13}C NMR (150.80 MHz, $CDCl_3$, TMS, ref. $CDCl_3$=77.0 ppm, selected data): 174.3, 71.9, 51.4, 37.4, 37.0 (t, J=18.2 Hz), 34.1, 31.9 (t, J=3.5 Hz), 29.5 (2 signals overlapped), 29.43, 29.41, 29.3 (t, J=3.7 Hz), 29.2, 29.1, 25.54, 25.51, 24.9, 22.7, 14.1. ESI-MS (ES$^+$): m/z = 319 (M+1).

Procedure B: in the external vessel of the apparatus shown in Figure 2, 210 mg of lithium wires finely cut and a little magnetic bar were introduced under argon. In the internal vessel, containing a little magnetic bar, 350 mg of transmethylated extract dissolved in methanol (5 mL) and 100 mg of PtO_2 were introduced. The apparatus was closed with a screw cap and magnetic stirring started: 3.5 mL of D_2O were then slowly added, using a syringe, to the lithium through the septum seal at the bottom of the apparatus (see Figure 2). At the end of the reaction (easily checked by the disappearance of the lithium), the apparatus was carefully disassembled, the catalyst was removed by filtration, and the solution was concentrated. As described in procedure A, the ^{1}H NMR analysis revealed the absence of vinyl hydrogen signals. The spectral data agree with those reported in procedure A.

(9*S*)-9-hydroxyoctadecanoic10,11,12,13-d_4 acid has been obtained as main deuterated product by hydrolysis with KOH/CH_3OH as previously reported [55] for the not-labelled analogue. ^{1}H NMR (400 MHz, $CDCl_3$): 3.64-3.54 (m, 1H, CHOH), 2.34 (t, 2H, J = 7.5 Hz, CH_2COO), 1.65–1.50 (m, 2H, CH_2CH_2COO), 1.50–1.15 (m, 21 H), 0.88 (t, 3H, J = 7.1 Hz, CH_3).^{13}C NMR (100.56 MHz, $CDCl_3$, TMS, ref. $CDCl_3$=77.0 ppm, selected data): 179.1, 72.0, 37.3, 37.2-36.7 (m), 33.9, 31.9, 29.4, 29.3, 29.2, 28.9, 25.5, 24.6, 22.7, 14.1. ESI-MS (ES^+): m/z = 327 (M^+ + Na).

CONCLUSION

The widespread occurrence of hydroxystearic acids in many different types of living organisms (animals, plants, fungi, yeasts, bacteria, *etc.*) makes difficult to find general statements to describe their biological role and functions. Instead, the purpose of this review is to summarise the usefulness that specifically deuterated HSAs have found in various research fields.

Generalisations may regard the deuteriation methods, which are restricted to few, giving both site specificity and acceptable chemical yield: then, depending on the particular kind of organism studied, the purpose of application of one, or more, deuterated HSAs, becomes different. For these reasons we give, as a conclusion of this paper, a tabular survey (see Table 1) of the deuterated mono- and di-hydroxystearic acids yet prepared and available, with the hope that this could suggest further investigations in other, not yet recognized, biological fields.

This review prompted us also to present a procedure which could be particularly suitable for the production of specifically deuterated hydroxystearic acids starting from unsaturated fatty acids.

Table I. Monohydroxy- and dihydroxy-deuterated stearic acids and derivatives

Hydroxyderivative	Ref.		Ref.
2-HSA-9,10-d_2 methyl ester	[20]	11-HSA-14,14-d_2 methyl ester	[42]
6-HSA-2,2-d_2 methyl ester	[31]	12-HSA-11-d acid	[47]
6-HSA-6-d methyl ester	[34]	12-HSA-12-d methyl ester	[34]
7-HSA-2,2-d_2 methyl ester	[31]	12-HSA-2,2-d_2 methyl ester	[31]
7-HSA-10,10-d_2 methyl ester	[42]	12-HSA-9,9-d_2 methyl ester	[42]
8-HSA-2,2-d_2 methyl ester	[31]	13-HSA-13-d acid	[48]
8-HSA-5,5-d_2 methyl ester	[42]	13-HSA-2,2-d_2 methyl ester	[31]
8-HSA-8-d methyl ester	[34]	13-HSA-16,16-d_2 methyl ester	[42]
8-HSA-11,11-d_2 methyl ester	[42]	14-HSA-14-d methyl ester	[34]
9-HSA-2,2-d_2 methyl ester	[31]	16-HSA-16-d methyl ester	[34]
9-HSA-9-d acid	[27]	17-HSA-17-d methyl ester	[36, 41]
9-HSA-9-d methyl ester	[27]	17-HSA-17,18,18-d_3 acid	[29]
10-HSA-2,2-d_2 methyl ester	[31]	17-HSA-16,16,18,18,18-d_5 methyl ester	[36, 41]
10-HSA-10-d acid	[29, 47]	18-HSA-18-d methyl ester	[29]
10-HSA-10-d methyl ester	[34, 46, 32]	9,10-di-HSA-8,8,11,11-d_4 acid	[52]
11-HSA-2,2-d_2 methyl ester	[31]	12,13-di-HSA-9,10-d_2 methyl ester	[49, 50, 51]
11-HSA-8,8-d_2 methyl ester	[42]	9-HSA-10,11,12,13-d_4 acid	This chapter
11-HSA-12-d acid	[47]	9-HSA-10,11,12,13-d_4 methyl ester	This chapter

REFERENCES

[1] McDermott, G. N. (1982). *Miscellaneous oil and fat products in Bailey's industrial oil and fat products,* Swern D. Ed., John Wiley & Sons: New York.

[2] Teeter, H. M., Gast, L. E., Bell, E. W. & Cowan, J. C. (1953). *Ind. Eng. Chem.,* 45, 1777–1779.

[3] Farroq, M., Ramli, A., Gul, S. & Muhammad, N. (2011). *J. Appl. Chem.*, 11, 1381–1385.

[4] Pryde, E. H., Princen, L. H. & Mukherjiee, K. D. (1981). *New sources of fats and oils*, American oil Chemists Society: Champaign.

[5] Svensson, M. (2010). Surfactants Based on Natural Fatty Acids, in *Surfactants from Renewable Resources* Kjellin M. and Johansson I., Eds., John Wiley & Sons: Chichester, UK.

[6] Petrović, Z. S., Ivetković, I., Hong, D., Wan, X., Zhang, W., Abraham, T. & Malsam, J. (2008). *J. Appl. Pol. Sci.*, 108, 1184–1190.

[7] Hamaguchi, T. & Atsuhito, M. (2006). Plasticizer for biodegradable resin, US Pat No. US20060276575.

[8] Naughton, F. C. (1974). *J Am. Oil Chem. Soc.*, 51, 65–71.

[9] Thetford, D. (2008). Paint Compositions, EP Patent No. 1, 383, 840.

[10] Marrion, A. R. (2004). Binders for conventional coatings in *The Chemistry and Physics of Coatings*, Marrion, Alistair R. and Marrion, A. Ed., RSC Publ: Cambridge, 96–150.

[11] Tadros, T. F., Dederen, C. & Taelman, M. C. (1997). *Parfuemerie und Kosmetik,* 78, 30–34.

[12] Drovetskaya, T. V., Yu, W. H., Diantonio, E. F. & Jordan, S. L. (2010). *Hair styling and conditioning personal care films,* US Pat No. US, 0247459 A1, 2010.

[13] Grissett, G. A., Keenan, D. M., Macedo, F. A. & Williams, D. R. (2008). Fibrous toilette article US Pat No. US 10/938,384.

[14] Schwab, W., Davidovich-Rikanati, R. & Lewinsohn, E. (2008). Plant J., 54, 712–732.

[15] Hughes, N. E., Marangoni, A. G., Wright, A. J., Rogers, M. A. & Rush, J. W. E. (2009). *Trends Food Sci. & Techn.*, 20, 470–480.

[16] Toro-Vazquez, Jorge F. & Morales-Rueda, J. (2010). *Food Biophys.*, 5, 193–202.

[17] Gunstone, F. D. (1996). *Fatty acid and lipid chemistry*, Springer Verlag: Berlin.

[18] Fulco, A. J. (1983). *Progr. Lipid Res.,* 22, 133–160.

[19] Harwood, J. L. & Russell, N. J. (1984). *Lipids in Plants and Microbes*, Allen and Unwin: London.

[20] Jenske, R. & Vetter, W. (2008). *J. Agric. Food Chem.*, 56, 5500-5505.

[21] Takatori T. (1996). *Forensic Sci. Int.*, 80, 49–61.

[22] Makristathis, A., Schwarzmeier, J., Mader, R. M., Varmuza, K., Simonitsch, I., Chavez, J. C., Platzer, W., Unterdorfer, H., Scheithauer, R., Derevianko, A. & Seidler, H. (2002). *J. Lipid Res*., 43, 2056–2061.

[23] Heng, L. Method for producing natural perfume γ -dodecalactone with microbial transformation Pat. No. CN, 2009-10065540, *Chem. Abstr.* 154, 267914.

[24] Kim, A. Y. (2005). Application of biotechnology to the production of natural flavour and fragrance chemicals In: Natural Flavors and fragrances: chemistry, analysis, and production, Frey, C, and Rouseff, R. L. Eds., *ACS symposium series 908*, ACS, 60–75.

[25] Wanikawa, A., Shoji, H., Hosoi, K. & Nakagawa, K.-I. (2002). *J. Am. Soc. Brewing. Chem.,* 60, 14–20.

[26] Calonghi, N., Cappadone, C., Pagnotta, E., Farruggia, G., Buontempo, F., Boga, C., Brusa, G. L., Santucci, M. A. & Masotti, L. (2004). *Biochem. Biophys. Res. Commun*., 314, 138–142.

[27] Calonghi, N., Cappadone, C., Pagnotta, E., Boga, C., Bertucci, C., Fiori, J., Tasco, G., Casadio, R. & Masotti. L. (2005). *J. Lipid Res.,* 46, 1596–1603.

[28] Cristofolini, L., Fontana, M. P., Boga, C. & Konovalov, O. (2005). *Langmuir,* 21, 11213–11219.

[29] Plettner, E., Slessor, K. N. & Winston, M. L. (1998). *Insect Biochem.* Molec. Biol., 28, 31-42.

[30] Badami, R. C. & Patil, K. B. (1981). *Prog. Lipid Res*., 19, 119–153 and ref. therein.

[31] Tulloch, A. P. (1978). *Org. Magn. Res*., 11, 109-115.

[32] Radhakrishnan, R., Robson, R. J., Takagari Y. & Khorana, H. G. (1981). *Meth. Enzym*., 72, 408-433.

[33] Bertucci, C., Hudaib, M., Boga, C., Calonghi, N., Cappadone, C. & Masotti, L. (2002). *Rapid Commun. Mass Spectrom*., 16, 859-864.

[34] Matsumori, N., Yasuda, T., Okazaki, H,; Suzuki, T., Yamaguchi, T., Tsuchikawa, H., Doi, M., Oishi, T. & Murata, M. (2012). *Biochem*., 51, 8363-8370.

[35] Lund, E., Budzikiewicz, H., Wilson, J. M. & Djerassi, C. (1963). *J. Am. Chem. Soc.,* 85, 1528-1534.

[36] Heinz, E., Tulloch, A. P. & Spencer, J. F. T. (1969). *J. Biol. Chem*., 244, 882-888.

[37] Mattson, B., Foster, W., Greimann, J., Hoette, N. L., Mirich, A., Wankum, S., Cabri, A., Reichenbacher, C. & Scwanke, E. (2013). *J. Chem.* Ed., 90, 613-619 and ref. therein.

[38] Polanyi, M. & Horiuti, J. (1934). *Trans. Faraday Soc.*, 30, 1164-1172.
[39] Koritala, S., Selke, E. & Dutton, H. J. (1973). *J. Am. Oil Chem. Soc.,* 50, 310-316.
[40] Crane, S. N., Bateman, K., Gagne, S. & Levesque, J.-F. (2006). *J. Labell.* Compd. Radiopharm., 49, 1273–1285.
[41] Tulloch, A. T. & Mazurek, M. (1963). *J. Chem Soc. Chem Commun.*, 692-693
[42] Tulloch A. P. (1976). *Lipids*, 12, 92-98.
[43] Hünig, S., Lücke, E. & Benzing, E. (1958). *Chem. Ber.,* 91, 129.
[44] Hünig, S. & Buysch, H. J. (1967). *Chem. Ber.*, 100, 4010.
[45] Bowmann, R. E. & Fordham, W. D. (1952). *J. Chem. Soc.*, 3945-3949.
[46] Yamaguchi, T., Suzuki, T., Yasuda, T., Oishi, T., Matsumori, N. & Murata, M. (2012). *Bioorg. Med. Chem.*, 20, 270-278.
[47] Rontani, J. F. & Aubert, C. (2008). *Amer. Soc. Mass Spectrom.*, 19, 66-75.
[48] Albrecht, W., Schwarz, M., Heidlas, J. & Tressl, R. (1992). *J. Org. Chem.,* 57, 1954-1956.
[49] Rakoff, H. & Emken, E. A. (1978). *J. Labell. Comp. Radiopharm.*, 15, 233-252.
[50] Rakoff, H. & Emken, E. A. (1982). *J. Labell. Comp. Radiopharm.*, 19, 19-33.
[51] Rakoff, H. & Emken, E. A. (1963). *J. Am. Oil Chem. Soc.*, 60, 546-552.
[52] Buist, P. H. & MacLean, D. B. (1980). *Can J. Chem.*, 59, 828-838.
[53] Gioacchini, A. M., Calonghi, N., Boga, C., Cappadone, C., Masotti, L., Roda, A. & Traldi, P. (1999). *Rapid Commun. Mass Spectr.*, 13, 1573-1579.
[54] Gioacchini, A. M., Pezzetta, D., Calonghi, N., Boga, C., Cappadone, C., Masotti, L. & Traldi, P. (2000). *Rapid Commun. Mass Spectrom.*, 14, 1954–1956.
[55] Parolin, C., Calonghi, N., Presta, E., Boga, C., Caruana, P., Naldi, M., Andrisano, V., Masotti, L. & Sartor, G. (2012). *Biochim. Biophys. Acta,* 1821, 1334–1340.
[56] Ebert, C., Felluga, F., Forzato, C., Foscato, M., Gardossi, L., Nitti, P., Pitacco, G., Boga, C., Caruana, P., Micheletti, G., Calonghi, N. & Masotti, L. (2012). *J. Mol. Catal. B: Enzymatic*, 83, 38–45.
[57] Smith, C. R., Wilson, T. L., Melvin, E. H. & Wolff, I. A. (1960). *J. Am. Oil Chem. Soc.,* 82, 1417–1421.

[58] Jensen, N. J., Tomer, K. B. & Gross, M. L. (1985). *Anal. Chem.*, 57, 2018-2021.

[59] Hesse, M., Meier, H. & Zeeh, B. (1987). *Spektroskopische Methoden in der Organische Chemie Thieme* Verlag: Stuttgard.

Editors' Contact Information

Dr. Yunfeng Lin, M.D., Ph.D., D.D.S.,
Professor of Department of Oral and Maxillofacial Surgery
Assistant Dean of West China College of Stomatology
Vice-director of State Key Laboratory of Oral Diseases
Sichuan University No. 14., 3rd Sec, Ren Min Nan Road
Chengdu, 610041. P. R. China
Tel: 86-28-85503487
Email: yunfenglin@scu.edu.cn
dentistlin@hotmail.com

Dr. Qiang Peng, Ph.D.
Assistant Professor of State Key Laboratory of Oral Diseases
Sichuan University No. 14., 3rd Sec, Ren Min Nan Road
Chengdu, 610041. P. R. China
Tel: 86-28-85503487
Email: lijm2002@163.com

INDEX

C

D

E

F

G

H

I

J

K

L

M

N

O

P

Q

R

S

T

U

V

W

X

Y

Z